CANADIAN INSTITUTE FOR
LAW AND POL
517 COLLEGE STREE
TORONTO, ONT.

MIRACULOUS LIFE CHAIN

The Essence of Evolution from the Universe to Mankind

Yoshiko Y. Nakano
President, OISCA-International, Tokyo

First published by Helix Editions Ltd., Centrepoint, Chapel Square, Deddington, Oxon OX15 0SG, United Kingdom.

Copyright © 1994 by Yoshiko Y. Nakano

All rights reserved. No part of this publication may be reproduced, stored in a retrieval system, or transmitted, in any form or by any means, electronic, mechanical, photocopying, recording or otherwise, without the prior permission in writing of both the copyright owner and the publishers of this book.

Translation by Tadashi Watanabe,
Akio Tabata, and Fumio Kitsuki

Edited by Norbert Netzer,
John Tomlinson, and Eric Waldbaum

Jacket design by Alan Forster

Printed and bound by Biddles Ltd., Guildford, Surrey, U.K.

British Library Cataloguing-in-Publication Data

A catalogue record for this book is available
from the British Library.

ISBN 1-898271-00-3

Table of Contents

Preface i
Introduction v
Chapter ONE: The Universe as the Origin of Life 1
   A visitor from space...1
   Scientific theories about the creation of the Universe 6
   Science should play within the wider laws of Nature 9
   Human beings need to choose wisely 11
   Results of observations by the Hubble space telescope ...12
   The Universe as the progenitor of all life 13
   The endlessly expanding Universe and "dark matter" 15
   The wonder of Earth 17
   Changes to Earth's ecosystem by human beings 21
   Serious global environmental issues 25
   The Moon and mankind 27
   From "Earth Ethics" to "Universe Ethics" 30

TABLE OF CONTENTS

Chapter TWO: Human Life is the Art of God 35
    Human beings as a mini-Universe 35
    Three functions of cells 36
    Poverty, population, and environmental problems 40
    The advances of genetic research 42
    Birth of life, a sacred event 44
    Research must be "humanly" motivated 49
    Activities of the ultra-microworld 51
    The new cosmology emerging from quantum physics ... 52
    The limitlessly expanding Universe 54
    Earth shares a common fate with the Sun 54
    Connection of spirit transcending space and time 55
    Return to Nature's law and order 56

Chapter THREE: Roots of the Japanese Outlook 59
    Culture closely linked with the forests 59
    Views on Nature nurtured by Japanese 61
    The impact of modern thought 65
    Kojiki informs about ancient Japanese thinking and life 65
    Poly-gods and mono-god 71
    Agriculture as *musubi* of Heaven and Earth 73

TABLE OF CONTENTS

Chapter FOUR: The Truth of Life Chain and Cycle 81
    The Universe is full of *'spiritons'* 81
    Flow of Life—Will surpassing human intellect 82
    Eternity of Life 83
    Selfishness and tyranny of human beings 84

Chapter FIVE: A New Earth Ethics 89
    The dawn of a new era 89
    Response to rising population and poverty 91
    Scientific advances 92
    Water, soil, and forest 95
    A lesson from the past 98
    An idea whose time has come 99

Appreciation

Index

Preface

Yoshiko Y. Nakano is President of the *Organization for Industrial, Spiritual and Cultural Advancement.* OISCA-International was founded on October 6, 1961. Today, she heads this important non-governmental organization, the first major international NGO to emerge in Japan.

Recognized for its long-standing efforts and contributions to sustainable development, it has dispatched thousands of volunteer experts to work in developing countries throughout Asia. OISCA-International received the "Earth Summit Award" from the United Nations in 1993.

OISCA-International operates from the conviction that sound development must be rooted foremost on agriculture and rural-based primary industries such as farming, fishing and forestry. While promoting no specific religion, OISCA-International stresses a spiritual component along with material development. Like a cart with two wheels, progress needs both knowledge and heart. Heart-to-heart communication is a cornerstone of OISCA-International. It gathers people who subscribe to its ideas and its spirit for service and international cooperation.

At its training facilities in Japan, more than 4,000 trainees have gained practical skills in technologies suited to rural development. In addition, thousands more have attended more than a score of training facilities in other Asia-Pacific countries. They have learned also improved methods for agriculture, fishing, and forestry through on-the-job training.

Wherever it works to foster rural community development, OISCA-International cooperates with partner agencies and indigenous organizations. It encourages and mobilizes volunteers, principally from Japan. Their projects have developed more vigorous rice crops in Karanganyar, Indonesia. A trilateral cooperation project with the Asian Development Bank and the Philippine government planted 250 hectares of mangroves in the Philippines.

An international *Love Green Campaign* carries on extensive tree planting activities to offset the destruction of forests. *Children's Forest Program* has planted hundreds of hectares of trees in countries where tree planting is badly needed. Children learn the value of tree planting and the importance of trees to life by working side-by-side with volunteers and through exchange programmes.

Membership in OISCA-International extends throughout the world to more than 100 countries. National affiliates of OISCA form chapters and each chapter may have many branches. There are affiliates of OISCA in Bangladesh, Brazil, Cambodia, Federated States of Micronesia, Fiji, Hong Kong, India, Indonesia, Iran, Israel, Kenya, Korea, Laos, Malaysia, Nepal, Pakistan, Palau, Papua New Guinea, Paraguay, Philippines, Singapore, Sri Lanka, Taiwan, Tanzania, Thailand, Uruguay, and Viet Nam. Collaborating organizations include private voluntary organizations in Ecuador, Germany, the United Kingdom, and the United States.

OISCA-International has organized a series of Asian-Pacific Youth Forums for Community Development, involving more than 20,000 youth leaders and workers. In cooperation with

United Nations agencies, national governments, and partner organizations, the Forums have included representatives from Australia, Bangladesh, Brunei, Cook Islands, Darussalam, Federated States of Micronesia, Fiji, Hong Kong, India, Indonesia, Japan, Kiribati, Malaysia, Marshall Islands, Nepal, Pakistan, Palau, Papua New Guinea, Philippines, Republic of Korea, Singapore, Solomon Islands, Sri Lanka, Taiwan, Thailand, Tonga, Tuvalu, Vanuatu, and Western Samoa.

Special training programmes for women advance OISCA's objectives of women-in-development. A model programme for women operates in Bangladesh. Under the auspices of the United Nations High Commissioner for Refugees, OISCA undertakes training for Tibetan youth in India, ranging from agriculture to tailoring.

Funding comes from a wide range of sources, including membership dues, donations, and grants. Supporters include the Asian Development Bank, the International Bank for Reconstruction and Development, the Economic and Social Commission for Asia and the Pacific of the United Nations, the Government of Japan, the Sasakawa Peace Foundation, and the Lions International Foundation.

Introduction

A few years ago I briefly described the philosophy and activities of the *Organization for Industrial, Spiritual and Cultural Advancement—International* in Asia and the Pacific countries in my book *A Message from OISCA to Mother Earth*. Born in Japan, OISCA-International has grown on the soil of Asia and attained 30 years of international cooperation activities, working hand-in-hand with the people of this geographically wide and diverse region.

OISCA's creed celebrates life and the wonders of life that can be experienced through observing the Universe and the human race, in both individual and collective relationships. Some may ask why OISCA—dedicated to international cooperation—deals with the Universe and life. I hope the reader will find the answer to that question in the text. OISCA's position on international cooperation and environmental issues derives from understanding the linkage from the Universe to Life. The thesis describes the fundamental aspect of the "Life chain and Cycle" that connects the Universe, human beings and all other living beings.

I have also touched on the connection between the views of Nature and the Universe held by Japanese people in ancient times and contemporary Japanese thought. Modern ideas and life rest on the heritage of our ancestors who viewed their lives as co-existing with Nature in harmony.

The International Foundation for Cultural Harmony, an affiliate of OISCA-International registered with the Ministry of Education, Japan, operates several astronomical observatories across the country. One of them, Moonlight Astronomical Observatory, stands on the foot of Mt. Fuji at a small community called Kannami almost in the middle of Japan's main island. Supported by well-wishers who love astronomy, it passed its 36th year, and is four years older than OISCA. Most of the references to astronomy, appearing in this thesis, are from the materials of the Moonlight Observatory. I am grateful for the cooperation of its staff and researchers.

The Observatory, equipped with a 500mm-diameter reflection telescope, a 200mm-diameter refraction telescope and some other optical equipment, is carrying out observations of celestial bodies. It has already succeeded in discovering as many as 159 asteroids, of which 17 have been officially registered with the International Asteroid Center at the Smithsonian Institute, USA. Already, six of the 17 have been given names beginning with "OISCA".

The Observatory is contributing positively to society and education. Such a contribution is its primary task. It houses on its premises a planetarium and an earth-science museum. The museum has more than a thousand fossils collected from many parts of the world.

Celestial observation sessions, study programmes and training courses often include overnight stays for primary and middle school pupils, students, youth, school teachers and adults at the quite extensive accommodation that is open to the public. Programmes are designed to advance knowledge through an understanding of astronomy and life on earth. In a broad sense, the Moonlight Astronomical Observatory is a center for environmental education.

INTRODUCTION

Children and adults alike are amazed to see the galaxy in which we live, the other planets, and the various realities of the celestial space, through the eyes of its optical equipment. Youngsters are often quite surprised to see and touch the stone-like big feces of dinosaurs at the earth-science museum. As we watch their lively faces, we realize the importance of our observatory's role in supplementing school education that is often inflexible.

Everybody ought to know the value of life. One of the most important aspects of learning is to recognize that life is not for oneself alone, but rather it is sustained in broad association with others. Life also is not merely for the current generation. It may be no exaggeration to say that the essence of environmental education is the promotion of the importance of these interrelationships. Too many modern inhabitants of earth, who are excessively concerned with their own claims to individual rights, know surprisingly little about their own life. The astronomical observatory is an ideal place to nurture a wider perspective and a richer sensitivity in the minds of children to help them in the future. At the same time it serves to demonstrate the realities of the Universe through science, astronomical study, and observation of celestial bodies. This nurtures in the human heart gratitude for life and moderation.

Even for a beginner, it is possible to learn through astronomical observation that the Universe is so expansive, so beautiful, so orderly. It is really an impressive sphere. One can also understand that our life—so small—is living on the accumulation of elements sent down from far away corners of the Universe where the birth and death of stars are ceaselessly occurring. Many think that our life is just a matter of course. It is important to know that Life is linked with the Universe.

The elements composing our physical bodies come from the Universe. Our life is a gift from the Universe. It was three thousand five hundred million years ago that life on our planet emerged in the sea. DNA, deoxyribonucleic acid, was already in the primitive cell. Since then, genetic information written in the DNA has been inherited through generations in conformity with the law of life. Surprisingly, the cells in our body have the same structure as the first life, transcending 3.5 billion years.

Human birth starts with the receipt of one spermatozoon by an ovum, then reproduces history, so to say, evolving from primitive forms in the ocean to become a human life. Thus, it sometimes seems that ontology recapitulates phylogeny.

Modern science and technology aim to produce a faster and more efficient world. Nonetheless, the drama of life that is played out in the mother's womb for 280 days has not been shortened or simplified. The laws of life that cause a child to be born progress in nine months through the long, long history of living things, which is said to be 3,500 million years. What better example could there be that mankind lives only in relation to other lives and is governed by the laws of Nature.

There are many who are concerned with miracles and the wondrous happenings around them. However, few are impressed by the real wonders—that everybody, without exception, has endlessly long and eternally strong connections with the past. Such negative attitudes may be because people's thoughts and spirits cannot see beyond the current materialistic way of thinking.

It seems to me that the Universe has evolved with a purpose—to give birth to Earth and ultimately to human beings. As Albert Einstein remarked "How much choice could God have when he created the Universe and the Earth?" His oft-quoted statement "[God] does not play dice" reflects a genuine feeling for such a purpose.[1] I am confident that the birth of the human race is not by chance, but is the inevitable result of the Will of the Universe.

The emergence today of serious environmental problems points to the need for equally serious reflection on human life. While it is true that we must pay more attention to protecting the environment and the lives of other bio-beings, the human species should not be ranked on the same level as other animals. While the egoistic and arrogant attitudes of people deserve to be blamed for environmental pollution and degradation, we must remember that human beings have an important mission bestowed by the Supreme Creator of the Universe that other organisms do not have. Annihilation and prosperity on Earth are equally in our hands.

The fate of Earth's ecosystem lies in our thinking and our actions. If we go wrong, we may cause the destruction of Mother Earth and even of the Universe. Bestowed with the highest intellect, coupled with the ability to speak and thus freedom to think and act, our human race has been able to expand its sphere of activities and develop technological capabilities beyond that of any other living beings. As a result of its brilliant advancements, the human race has gained the illusion that it has the power to govern and control everything. It is this arrogant and egoistic assumption that has led to the global crisis visible in today's environment problems.

---

[1] Einstein, Albert, letter to Max Born, 4 Dec 1926, in *Einstein und Born, Briefwechsel* (1969) p. 130. Jedenfalls bin ich überzeugt, dass er nicht würfelt. At any rate, I am convinced that He [God] does not play dice.

It is important to see life in perspective from its very beginnings if we wish life to continue. The benefits we receive from air, water, soil and the natural environment, as well as from fauna and flora that exist within the natural environment, are enormous and we must ensure that they will continue in years to come. The life that is bestowed on us is not for us alone, but must be shared in order to sustain its flow. What better expression for gratitude? For life to continue, sharing and gratitude must become a central and important part of our education.

When the mind actually grasps the benefits that the human race is receiving in a variety of forms, visibly and invisibly, from the environment and ecosystem—all originating from Mother Earth—it is a natural human instinct to feel gratitude and to want to put something back as a thanksgiving for having received these benefits. A sense of respect and generosity will gush out from the heart, leading to an adjustment of life style. Voluntary action can then conform with the flow of the Universe.

This applies to both countries and individuals. International cooperation should not be a one-way flow, nor should a reward be expected. The spirit of voluntary action should exist in all areas of international cooperation. Japan for instance, has been able to develop to its current state thanks to the supply of raw materials from the countries of the South. If continuing progress in development and the continuing supply of raw materials is the wish of the Japanese people, the flow must be returned to the people in the South. How should this be done? One way is to seek out and satisfy the needs of the people of the South. This requires genuine motivation and returns the gift of life to them. It is with this fundamental spirit that OISCA has been carrying out international cooperation activities with full enthusiasm and determination.

INTRODUCTION

When I visit various countries and meet with people of diverse cultures, I often come across circumstances and environments within which I feel a oneness, sometimes affection, even passion, without malaise at all. In particular, I am deeply touched when I experience other people's consciousness of Nature and their particular views of eternal life and of the Omnipotent Being. These insights sometimes come when I experience or observe the festivals and ceremonies that they organize. At other times, they come when I begin to understand how they treat their parents, land, mountains, and natural environment.

There are some who may view this as fuzzy thinking. However, it is typically Japanese to relate the understandings of others to the way they treat their forests, land, mountains and the natural environment. We respect those who behave with full dignity or with pure simplicity.

Sometimes I think if I were born in an Islamic country, I would have become a Moslem or if I had been born in a Christian country, I would today be a Christian. Similarly, I might have become a Buddhist, a Hindu, or a Jew. I might have become a member of whatever my native community associated with its culture.

Whatever the means of access to the truth may be, and however the methods of teaching may differ, the ultimate concern of human beings must be with Life itself. The goal, therefore, must be to achieve human behavior that is obedient to the Will of the Universe.

Since time immemorial, the Japanese have a history of respecting forests and paying profound respect to Nature. Through their life style, they developed a worshipful mind. They have nurtured their own ethnic views of the World, the Universe and of Life itself. Japan's traditions and cultures have been inherited through generations. Since they strived to live in harmony with Nature, our Japanese forebearers were able to pass on these traditions. They held a view of Nature that acknowledged the existence of spirit in trees, mountains, water, rivers, and stones. They developed a consciousness to live in conformity with the link and cycle of these lives.

Having been born in Japan, it is quite natural that I take the basic philosophy of *shinto*, the Japanese ethical religion, as an entrance to the truth. I picture in my mind a World where all Nature is respected and the Universe, Earth, the human race, and all other creatures live in grand harmony. Such a World is the symbol of eternal peace. I have chosen international cooperation as the positive and practical means to attain such a World.

A flow of money and materials is not the only means of international cooperation. Heart-to-heart symphony is its true means. Positive exchange programmes can generate mutual respect and trust. Learning about and from each other, teaching each other, helping one another, and enlightening each other are at the very center of activities that engender international cooperation A common bond based on the life chain needs to circulate around the world to make it stronger and richer. It should be the mission of the human race to create a World where all other bio-beings can fulfill their distinctive missions. This, of course, requires an understanding of the roles and distinctive missions of every other bio-being and a full appreciation of the heart of the omnipotent parent. True affluence and prosperity will only be found in such pursuit. In contrast to the dark destiny that currently overshadows Earth, I earnestly hope that a day will soon come when all the peoples of the world will be released from uncertainty and fear and live in harmony, based on their appreciation for the life they have been given. All living beings on Mother Earth can be linked with the fuller Life of the Universe. When the rhythmical sounds of such a grand symphony orchestra spread, in the heaven and on land as well as in the Universe, we can take it as the fulfillment of Life—gifted on Earth by the Will of the Universe.

I have devoted this book to my inspirations, thoughts and philosophy. While these guide the activities of OISCA-International, I have not attempted to address here the honest volunteer work that centers principally in Asia and the Pacific. My earlier book, *A Message from OISCA to Mother Earth*, contains a record of OISCA's activities in the region.

CHAPTER ONE

The Universe as the Origin of Life

A visitor from space

Sometime in the beginning of December 1992, an asteroid called Toutatis, estimated to be about one kilometer in diameter, passed very close to our planet. Very close was as far as ten times the distance between Earth and the Moon. Certain scientists warned that it might again enter into Earth's gravitational field in September 2004. Then, according to calculations by some specialists, its distance from our planet would be 1.56 million kilometers, that is, still four times as far as the distance between the Moon and Earth—380,000 kilometers. Thus, the possibility of its collision with our planet is almost none, or at least, implausible.

A similar asteroid—or perhaps a meteorite—slightly larger than Toutatis is thought to have collided with Earth about 65 million years ago. The weather became freakish and changes due to dust saturation in the atmosphere led to the total extinction of dinosaurs. This hypothesis now seems to be widely confirmed. A release by NASA, the National Aeronautics and Space Administration of the United States, assesses the possibility of an asteroid's collision with Earth in our life-time as one in ten-thousand. Should this occur, human civilization might follow the same fate as the dinosaurs. The virtual "dinosaurs" on Jupiter may well have faced a similar prospect in July 1994 when the 20-piece "Shoemaker-Levy" comet collided with the planet.

It is not my intention to speculate on the annihilation of human beings. Rather, I am deeply impressed by the advancements of space science. Science appears to be uncovering, one after the other, previously unknown realities of the Universe. Scientific investigations are revealing the close relationship between activities of the Universe, Earth, and the lives of human and all other living beings. Deep impressions made by these revelations have led me to the realization that mankind has an important mission to fulfill to sustain the life of all in the Universe.

With this picture in mind, I would like to introduce the idea of the miraculous Life Chain, a linkage of the Life Cycle with the Universe. I contemplate this in my daily life. It helps in my understanding of the value of life and helps to reveal its mystery. If we want to know the reality of our own life, we must search beyond the things on our planet. The origin of life—as the origin of the Universe—lies in the Universe.

International Space Year with the theme "Mission to Planet Earth" was designated by the United Nations and observed in 1992. Since each member nation is composed of its people, it can be said that the decisions of the United Nations reflect the will of all of us, even if one cannot claim that it always represents the unanimous wish of the people. I am unaware of any serious objection to the UN's designation of International Space Year. For perhaps the first time in the history of the UN, the International Space Year raised its focus to the Universe and allowed the struggles and antagonisms between and among races on Earth to lose significance. Taking the perspective of the Universe allows us to observe our own planet, to see what is life-sustaining from the vantage point of outer space.

There are said to be more than 200 billion galaxies in the vast Universe—according to some theory, perhaps more than 1,000 billion. Among the galaxies, there are many that are larger than our own. We learn in school that light travels a distance equal to seven and a half times around the Earth in a second. We also learn that the speed of light is constant under any circumstances, about 300,000 kilometers per second. Measured by the speed of light, the diameter of our galaxy is said to be about 100,000 light-years. Just as the gymnasium of a school is tiny compared to Earth, our planet is tiny compared to the galaxy, much less to the vast Universe.

Our solar system, with the Sun in the center and Pluto at the extreme outside circle, extends for some five and a half hours as light travels. The solar system is but a very small body in the community of galaxies. In scale, it is smaller than the nucleus of a cell in a human body that is made up of more than 60 trillion cells. Viewed in this manner, each human is close to "nothing," much smaller even than an atom as compared to a human body.

When we accept that a human—who is close to nothing—is connected with the Universe through a chain, invisible to our naked eyes, transcending light years of space and time, we realize the fantastic mystery of the Life Chain. Life is so dignifying.

With my conclusion now stated, I would like to explore with you the insights into this infinite linkage of life that brought me to my conclusion.

Quite a large number of astronauts already have flown in space. Looking down from space, these astronauts experienced such feelings of admiration and love that they felt compelled to express them as messages to Earth. The space where these astronauts have flown is only 300-400 kilometers above the ground. It is just like a thin surface of Earth's sphere. Yet, such a tiny distance is wide enough for us to realize that Earth is the only house for mankind.

In 1969, America's Apollo astronauts landed on the Moon. This was a remarkable advancement. One of the astronauts remarked:

> For mankind living on Earth, the surface of the Moon was nothing but a dead environment. Life outside an atmospheric sphere might bring forth an evolution that may occur once in several billion years, as occurred several billion years ago when organisms from the young sea alone landed on the ground and developed a variety of means to live in what was then a dead environment. Stronger species may emerge in severe environments.[2]

As we look back on the history and evolution of living beings, this observation deserves our concern.

Mini-stars repeatedly crash into one another in space and break into smaller pieces. Leaving their regular orbit, some of them may strike Earth. They are called meteorites. More than two thousand meteorites so far have been found throughout the world. In addition, there may be yet undiscovered ones in the Antarctic, forests, deserts, and even at the bottoms of deep seas.

Fortunately, the asteroid Toutatis passed our planet without any harm to our life. Soon thereafter, however, Japan had another visitor—a small meteorite. On December 10, 1992,

---

[2] Tachibana, Takashi, *Return from the Universe*, Chuo Koron Publishing Co. (Tokyo:1983), p. 359-361. The author has summarized the interview with the astronaut.

while Toutatis was capturing the attention of the mass media, this meteorite fell down on a private house at a small town called Mihozeki in Shimane Prefecture, southwest of Japan's main island. The "Mihozeki Meteorite" was estimated to be 4,600 million years old. It was about 4.6 billion years ago when our Earth was created. So the Mihozeki Meteorite is of similar age to Earth. It is perhaps because of this similarity in age—making us comrades since the time of the formulation of the nebulae in the primitive sun—that a small meteorite, only a few kilograms in weight, attracts our attention.

It is said that this meteorite drifted in space without colliding with any other body for 61 million years before it reached Earth in 1992. The shooting stars we sometimes see as we watch the sky at night are the burning-out phenomena of meteorites and space dust as they enter Earth's atmosphere. We can appreciate that the atmosphere protects life on Earth not only from ultraviolet rays harmful to human and other beings, but also from such materials from space. If this important protective role of the atmosphere were absent, tremendous numbers of meteorites from within the entire vastness of space—those as small as stones as well as larger ones—would collide with our life sphere.

Think of the surface of the Moon covered with large and small craters. The pocked surface indicates many collisions with meteorites. Since air does not exist on the Moon, craters maintain their original shapes and are not weathered—a clear difference with Earth. That many meteorites have collided with the Moon has been confirmed by analysis of Moon rocks. Researchers have found that the Moon's surface is covered by sands as deep as three meters. This sand or broken surface may well indicate the extent to which the Moon's surface has been crushed by meteorites.

If Earth was not protected by its atmosphere, it might have experienced a similar fate to the Moon. We can easily understand the importance of the role of air in these events. This suggests to us that we should have a wider appreciation for the value of air.

An unexpected visitor from space in the International Space Year has opened our eyes a bit wider to the nature of the Universe.

Scientific theories about the creation of the Universe

There were new discoveries one after the other in 1992. A particularly significant discovery about cosmic background radiation was made by a NASA survey satellite. Known as COBE (COsmic Background Explorer), this satellite was sent up in November 1989 and its historic discovery was announced in April 1992. The satellite detected radiation "fluctuations" in the cosmic microwave background about three 100-thousandth °K (absolute temperature) or 0.001 percentage in ratio.

Today, the generally accepted model of the creation of the Universe is the "big bang theory". This model posits that the Universe was born some 15 billion years ago by a great explosion. The basic idea of this model was first put forward in 1948 by Dr. George Gamow, a Russian-born American.

Was there no seed for the explosion? What was the shape before the Universe was born? Nobody knows the shape of 15 billion years ago. It is possible that there were fluctuations so that a seed would germinate. Because of this, the big bang is sometimes called God's blow.

Cosmic microwave background radiation was discovered in 1964. It is considered to be a trace of the big bang. Its temperature of about three degrees above absolute zero, or about minus 270° C, is believed to be spreading evenly in all directions of the cosmos. Even if this is really so, the big bang theory does not fully account for the origin of the Universe. Without "fluctuations," the Universe can neither expand nor contract. Around us we can observe similar phenomena, for instance the "calm" weather that often occurs and when we feel as if all living beings have stopped breathing.

Science is making further and rapid advances beyond the big bang theory. It seeks to clarify, theoretically, the mechanism for the creation of the Universe. Previously, this subject was thought to be the sanctuary or sphere of God. In this process has emerged the "inflationary model" first advanced by the U.S. scientist Alan Guth. "Inflationary" means that the Universe initially expanded at an increasing rate rather than the decreasing rate that we see today. The "inflationary model" includes an explanation for points that the big bang Universe model could not describe—namely, why the Universe became a fireball and why it began to expand. The model explains that, immediately after its birth, the Universe rapidly expanded with energy formed by the vacuum. Within a few moments of the big bang, the Universe was a dense fireball of elementary particles. The extraordinarily fast expansion caused the Universe to become uniform at all points of space and in all directions.

The "inflationary model" of the Universe led to the prediction that small density perturbations would be generated in the expanding Universe. At last, the fact has been discovered that very tiny fluctuations existed in the far remote past, the time of the birth of the Universe. Density perturbations explain how galaxies are created. Indeed, these fluctuations are considered to be the seeds of cosmology. A difference of three 100-thousandth of a degree is nothing for our everyday life. It does have very important meaning for the Universe fifteen thousand million years old. If the fluctuations did not exist, neither would the Sun, Earth, and Moon. Human beings would not have been born. We can believe that the origin of the life of the Universe lies with the fluctuations.

There are, of course, many models for the creation of the Universe. One advocating the formation of the Universe from "emptiness" woos my interest. Many scientists reacted in shock when Alexander Vilenkin, a Russian-born American, announced his thesis, based on intuition and observation. For some scientists, such "emptiness" was inconsistent with physical laws. The thesis was viewed as not scientific, applying only to philosophy or religion. Vilenkin explained that

> Far from connoting 'nothing,' a vacuum to the physicist is a state of minimum energy obtained in the absence of all particles.[3]

After Vilenkin first presented his theory, British physicist Stephen Hawking introduced a scenario to deal with the formation of the Universe, inflation, expansion, contraction, and extinction using quantum theory. He described it by means of mathematics. His scenario coincided with that of Alexander Vilenkin as the most likely process of evolution of the Universe. The theory is so complex that its details are difficult for those of us who are neither physicists nor mathematicians to follow. Scientists are introducing new

---

[3] Vilenkin, Alexander, "Cosmic Strings," *Scientific American*, v257, December 1987, p 94.

theories, one after the other, on the birth and life of the Universe, and there are various scenarios that can be proven by observations and mathematics.

However, contradictions are gushing out one after the other. Often, the same scientists reverse their own scenarios after new discoveries. It is really true that the Universe is not as simple as scientists—much less we non-scientists—might have perceived.

That modern science has begun to theorize about the creation of the Universe is an important development. Only with the theory of relativity, quantum mechanics and so forth can it be said that modern science has at last entered into what was considered to be the domain of philosophy and religion. Yet, the theories of the creation of the Universe are within the grasp of our science and logic, not beyond them. So long as observations depend on light or electromagnetic waves, large scale telescopes can discover data about the origin of the Universe but will not capture its essence. After all, today it is not possible to observe the Universe beyond 15 billion years ago, the origin of the Universe as defined by Hubble's theory.

Science should play within the wider laws of Nature

I have introduced the activities of science dealing with the origin of the Universe because OISCA's own philosophy carries a universal viewpoint. It advocates that human beings—along with the other creatures—are part of the infinite continuity of life that originated in the Universe. OISCA philosophy teaches that peace, prosperity and happiness are attained where all lives are connected and perpetuated in harmony. Of course, ours is a spiritual approach that differs from that of science. It is the difference of approaching from within rather than from outside. OISCA never denies the approach from outside. That is why we are impressed by the beauty and greatness of the life of the Universe as observed by such physical means as optical instruments. We pay our sincere respects and admiration to scientific efforts to search out continuously the physical and further realities of the Universe. Modern science has led human society to make great advancements. The eyes of science observe the cosmos in its materiality. Science and its enthusiasm are wonderful indeed. Without a spiritual understanding, however, science may not have a sound foundation.

Human beings place supreme trust in their own abilities to exploit all aspects of the cosmos through science. It generates a haughty mind. When scientists face the Universe with a haughty mind, their approach will ultimately lead to conflict with the Will of the Universe. Science and scientist, alike, live within the wider laws of Nature that are the manifestations of the Will of the Universe. Even if human beings, through the advancements of scientific technology, should be able to live on the Moon, they cannot avoid the law of the Universe.

Human attempts to advance cannot be sustained outside the wider laws of Nature. Where human beings lose their sense of moderation and develop conceit, they will inevitably meet unexpected turmoil along the road. The force and the consistency of the Universe never change. Humans should not be too absorbed by their own fantasies. Genuine and lasting development can only progress if it is in tune and flowing with the laws of the Universe.

It is important for human beings to have humility. If scientists pursue their research with a faithful mind and pursue a greater understanding of the real nature of the

Universe and the Omnipotent Being, then the essence of the Universe will come close and will help them encounter truth.

Human beings need to choose wisely

The Universe was born 15 thousand million years ago, a far more remote past than we can readily comprehend. Yet human intelligence is perhaps capable of measuring time back to the big bang. All beings, including human beings, live lives that have roots in the origin of the Universe. Human life is certainly a part of the flow of life within the Universe. Yet, no human witnessed, saw, or heard the creation of the Universe. When people pay respect to its dignity and offer gratitude to what they understand to be the universal, the omnipotent Founder of Life, no other can deny as non-scientific or irrational their confidence in the circumstances of creation.

What would happen if materialism alone dominated the minds of all people and conquered their hearts as it has in all too many cases? Hearts of no warmth would cause havoc within human society. Communities and cultures would then collapse more quickly even than physical desertification is extending across our planet. The objective of science is to serve human welfare, and to promote progress and happiness for a better global human society. Human beings must choose their path wisely if we are to achieve this objective in harmony with the Universe.

It is a welcoming tendency that science is embracing more comprehensive approaches. Earlier science was subdividing and polarizing. Present-day science must address a comprehensive framework and deal with life in total harmony. The outlook for integrated approaches is favorable. "Life" never exists in isolation. It is becoming more widely recognized that life interrelates visibly and invisibly, both here on Earth and with all of the Universe as well.

What are happiness and wealth for man? What is human life? What is Life itself? We are living in an age that must consider such issues, not only in education, but also in politics, economics, diplomacy, and international relations. The founders of modern science expressed deeply human convictions. Many among them had deep access to philosophy, religion, music, and art, all of which involve inner human qualities. In the interim, specialization has led to narrower understandings. The unacceptable results of such narrower understandings are now leading to a demand for a return to scientific thinking within the broader universal framework.

Results of observations by the Hubble space telescope

Let me continue to talk about space. The furthest space we can observe today is the cluster of galaxies about 10 billion light-years away. It was discovered in early December 1992 by America's Hubble space telescope, known as the "flying astronomical observatory". Launched in April 1990, Hubble carries the name of the famous astronomer Edwin Hubble who proved by means of observation that the Universe is expanding. The resolution of this telescope is so precise that a small coin can be identified at some 200 km distance. Since its view is not affected by the atmosphere, it can send down to Earth clear and beautiful cosmic images that no telescopes on the ground can achieve.

The Hubble telescope has found the formation of a "primitive star" in the Orion clouds some 1,500 light-years away from our planet. It has also discovered the phenomenon of a "black hole," the death of fixed stars, at the Virgo galaxy cluster about 59 million light-years away from Earth. These discoveries have confirmed the existence of black holes and primitive stars. They teach us, at the same time, that there is a "Cycle of Life" in the Universe, as well as on Earth. New stars are born and die. Our Life Cycle on Earth is in the eternal flow of the Universe.

In February 1993, Space Science Institute, Japan sent up the X-ray satellite "ASTRO-D," with the main objective of searching the events that occurred immediately after the birth of the Universe by observing space 10 billion light-years away and beyond. It is a remarkable advancement of science. After having orbited Earth, the satellite was named "Asuka," after a place that once housed an ancient capital. With good timing, as if arrangements for observation awaited, a supernova "SN1993J" appeared in the galaxy "M81" some 10 million light-years away from our planet. This event is said to be the explosion of a giant star that has come to the end of its life. Asuka's discovery of the scene immediately after the birth of a supernova attracted international attention.

The Universe as the progenitor of all life

The Universe repeats creation and extinction. This applies not only to galaxies and fixed stars and to life on Earth, but even to atoms. Elements on Earth were absorbed by solar system nebulae after having changed, within the galaxy, from gas to stars and vice versa. According to the big bang theory, the lightest elements, hydrogen and helium, were created first. They were followed by heavier elements such as carbon and oxygen, created one after the other in stars. Heavy elements spread in space through explosions of supernovae. They become gas, which then contract to form stars. Countless repetition of this process increases the volume of heavy elements in the galaxy. Old elements from repetitions of this process mixed with primitive ones from the early stage of the Universe. Our galaxy was born some 10 billion years ago. With the life expectancy of fixed stars averaging tens of millions of years, there should be stars with ancestors further than two hundred generations back. (The life expectancy of a star depends on its mass, the smaller the mass, the longer its life. It is said that the life expectancy of the Sun is 10,000 million years, while a star with ten times the mass of the Sun has a life expectancy of 30 million years.)

Elements that have a variety of bio-data exist in the human body, and on Earth in layers. Beyond the Sun exists an antecedent that gave birth to the Sun. All the elements that form our physical bodies, too, are parts of stars burnt out millions or billion of years ago, and created through a repeated process of explosions and contractions of stars.

It is only in the last ten years or so that the connection between the micro-world of elementary particles and the macro-world of the Universe has been clarified.

Judging from the elements of the body's composition, the origin of human beings lies in the Universe. Life and Earth are both given by the Universe, the parent. In the micro-world, too, the same principle applies to atoms, neutrons, protons, and even to elementary particles.

As one realizes scientifically the flow of elementary particles, we see that life exists as linkages and cycles from a far away and old Universe. From a spiritual point of view, human beings are a part of the flow of life originating in the remote Universe.

Is any life too small, too vulnerable to protect, if it is linked phylogenetically with the life of the Universe? If one thinks that human beings exist merely as material beings—

unconnected with the flow and the cycles that link mankind with the Universe—then, indeed, human and all other beings will become orphans.

The endlessly expanding Universe and "dark matter"

"Star burst" has also been discovered. It is a phenomenon of the birth of stars as a result of collisions among galaxies. Tens and even hundreds of thousands of big stars having more than 10 times as much mass as the Sun are born at a time. It is really surprising to know that 6,000 galaxies with such star burst phenomena have already been discovered. Star burst is something similar to the population explosion on our planet, particularly in the developing countries. In developing countries, as a result of the population explosion, hunger and poverty are dramatically increasing. Unlike our planet, it seems that the Universe can endlessly expand. It has the capacity to accommodate star bursts.

There was one more important discovery in 1992, namely the existence in the cosmic world of "dark matter." As it does not discharge light, radio waves or X rays, no telescopes can view it. Dark matter is invisible. The combined mass of matter visible through telescopes, such as galaxies, fixed stars and gas clouds is merely one-tenth of the mass of the Universe. What about the remaining 90 percent? Known as "lost mass," its presence was predicted by astronomers during the 1930's.

The reality of the presence of dark matter was almost confirmed by NASA's observation satellite. X-rays of gas clouds of 10 million° K in absolute temperature, showed that gas that normally scatters because of high temperature did not do so. As we experience in our daily life, gas scatters in our atmosphere unless restrained by a container or the like.

For instance, the surface temperature of the Sun is known to be about 6,000° K energy. This affects the life sphere of Earth from a distance of 150 million kilometers. The energy that reaches Earth is only one out of 2,200 millionths of the whole that the Sun dispatches. Even then, we are not fully using the Sun's blessing. Scientists think that 10 million° K temperature gas does not scatter in the endless Universe because it is pulled back by the gravity of invisible matter having 500 billion times the mass of the Sun. The strength of the gravitational force is far beyond our imagination.

If this is true, it seems to me that this invisible "dark matter" is supporting the life of the Universe. It is similar to air that has no smell, no color and cannot be felt by human eyes and skin; yet, all the while, it is protecting the life sphere of Earth. Chemically, air is composed of nitrogen, oxygen, carbon dioxide and other minor elements. Above the sky is stratosphere and above that lies the ozone layer. They are protecting our life sphere.

Dark matter cannot be identified by the eyes of telescopes. However, one may well be able to say that it sustains the life of the Universe. Scientific research is gradually proving the "link" in universality between our life sphere and the Universe.

The "dark" of "dark matter" may stem from the darkness of space itself. Although the advance of scientific technology has made it possible to observe groups of stars as far away as 10 billion light-years, surprisingly there has been no news yet that "air" such as that which surrounds our planet has been discovered in space.

Dark matter is not black, just as black holes are considered not exactly black. Dark matter must be colorless, transparent, odorless and clean. Like air, we do not feel its presence.

While it may lie outside the ability of our sensory apparatus to perceive it, nevertheless, it has tremendous value. I may describe it as the "formless being".

The wonder of Earth

Air exists on and around Earth to enable humans and all other living beings to sustain a life. Is this just an accidental phenomenon? It is amazing to know how many accidents have accumulated on Earth to create the only aqua-planet in the Universe. These are not merely accidental phenomena. It is reasonable to think that the Will of the Universe has a firm objective to have concentrated creative functions on Earth. It is reasonable, too, to believe that the Universe and Earth, as well as human beings and other creatures, are linked by a connection transcending time and space. I describe it as the miraculous Life Chain.

Where has life on Earth come from? The "origin of life" advocated by Aleksander Oparin sometime ago described it as arising solely within Earth's environment. As researchers have made advances, and elements and particles that become the roots of life have been discovered in space, it is now recognized that life can emerge anywhere provided that the necessary conditions are fulfilled. Very high grade organic particles are found in meteorites. The existence of amino acids, too, has been indicated. Materials having unique characteristics typical of life have been found in very small particles. Explanations of the origin of life do not have to be limited to Earth. The possibility of life in the Universe is increasing. Nevertheless, it is really amazing that in the vast Universe that is expanding at critical rate, the place where life was born is Earth.

How then was Earth born and how has it evolved? Meteorites and craters made by meteorites' collision with Earth millions of years ago offer evidence. It is possible to draw a scenario of the birth of Earth by analyzing the ingredients of meteorites and craters. By observing the crater-covered surface of the Moon and Mars, and the craters remaining on Earth's surface, one can imagine what the primitive Earth looked like.

As noted earlier, the Universe is repeating creation and extinction. Heavy elements spread across space by the explosions of supernovae. They become gas and then contract to form stars. Earth, too, might have followed a process similar to one of the stars. First, gas and dust spread across space by supernova explosions began to make a vortex, and a primitive sun was created by the concentration of the majority of gas in the center of the vortex. Left-out gas, dust and debris joined around the primitive sun to form mini planets, about 10 km in diameter, that then begin colliding. Some were crushed to pieces. Others merged. Through repeating these processes, primitive planets, as big as the Moon, were created. Planets having a greater gravity have a stronger attractive force and absorb mini-planets one after the other. Thus, our planet may have been born.

The primitive Earth, with acceleration of gravitational force, continued to expand while incorporating mini-planets. As it grew, its gravitational force became stronger, and colliding speed advanced. Powerful collisions melted at once the surface of the primitive Earth, and melted magma covered Earth's surface, on which additional collisions occurred, leading finally to an ocean of magma that covered the entire surface of the primitive Earth.

Mini planets, too, melted at the moment of collision and their ingredients separated as a matter of course. Heavy ones sank down, lighter gas formed atmospheric air, and water ran away as water vapor. A thick layer of water vapor atmosphere formed around the primitive Earth. The number of mini planets decreased slowly. As collisions ended, the atmosphere began to cool down. When the surface temperature reached 300 degrees, it began to rain, the first rainfall on the primitive Earth. Heavy rain continued almost endlessly. It was perhaps a greater flood than Noah's, far beyond our imagination. Ocean emerged on the primitive Earth. If Earth had been created a little closer to the Sun, the water vapor might have been spread into vast space and our planet might have become a oceanless planet like Venus.

Sunlight began to reach Earth's ground when the thick water-vapor-layer clouds ceased. Circulation of water, namely evaporation of ocean water and rain fall, began to take place. Then continents formed. Carbon-dioxide in the atmosphere decreased slowly as it was absorbed by oceans and continents. Various complex factors mixed. Then a miracle took place in the primitive ocean—the birth of life. It is clear today that primitive bacteria, a kind of algae, existed at least 3.5 billion years ago, which is proven by the existence of stromatolites, a fossil of prokaryote, discovered on the coast and in the interior of Western Australia. Through photosynthesis of these primitive bacteria, the ingredients of the atmosphere began to change.

Fossils of the first ground plants, called psilotinae as they look like psilophyta, were discovered in Canada and England. These plants appeared about 400 million years ago. For life to emerge from the ocean to land, atmospheric conditions must be in order. From these fossils, it is estimated that the volume of oxygen in the atmosphere had reached almost the same level then as today, and the ozone layer had been formed to protect life from harmful ultraviolet rays.

These primitive plants made soil on the rocky land, stored water, provided grazing for animals and grew plants that followed them. Repetition of these processes resulted in Earth's environment as we find it today. Only through the grand circulation of air and water have all living beings been able to evolve.

A Japanese planet scientist Takafumi Matsui writes in his book on cosmography,

> As we trace the 4,600 million-year history of Earth, it seems as if it has evolved wholeheartedly right from the beginning of its evolution to bring forth a sound environment for life. Oceans, lands and atmosphere were carefully controlled to allow life to be born. Once life is born, Earth involves the former as an important factor to maintain its own environment....[4]

Earth is keeping a regular, sound environment by means of breathing on its surface area, namely absorbing carbon dioxide and discharging it. Indeed, Earth has a careful feedback mechanism in order to maintain a sound surface environment for itself. Such mechanism is handled by the invisible hands of Mother Earth, where, too, we can witness the indefatigable Will of Earth to protect life.

These descriptions suggest that the birth of Earth is inevitable and very much so is the birth of human beings on Earth.

Human beings, together with animals and plants, are managing life while interrelating with the blessings of soil, water, air and solar energy. The basis of their eco-style lies in

---

[4] Matsui, Takafumi, *Report on the Universe*, Tokuma Shoten Publishing Co. (Tokyo:1993), p. 295-6.

the life activity of Mother Earth, involving the functions of the atmosphere, oceans, floors of the oceans, continents, and mantle.

As in earlier times, the Earth's present environment is sustained through a variety of Nature's functions. From this, we must know that when a piece of the chain is cut off, the balance will collapse, leading to the destruction of the environment.

Consider an organism. It lives in a food chain system, namely one eats others or is eaten by them. Small fish eat plankton. They are run after by medium size fish, which are run after by big ones, that are caught by upland animals. The food chain is not a one-way system. Exhausted big fish are eaten by plankton. Such about-face functions do exist. Complex interrelations are spreading throughout the organism. It is said that the ecosystem represents these exquisite structures that act in harmony and sustain conditions. So, if a part becomes out-of-order, the whole system will be affected adversely. The ecosystem works as long as its self-recovery function is intact. It proves to be exceptionally vulnerable when circumstances surpass its ability for self-recovery.

Changes to Earth's ecosystem by human beings

In the beginning, human beings, the latest newcomers to the natural ecosystem that was built on Earth over very long years, lived in wonderful equilibrium with Nature. In the course of time, they learned the means to produce food according to their needs. They also improved animal hunting skills. Through these means, they have been able to secure enough food.

Slowly, life expectancy expanded. Food production was further increased. As a result, the natural ecosystem became an inconvenient system for human beings. In order to support an ever increasing population, they began to loot Nature, making themselves, in the course of time, unconscious invaders, agents to confuse the natural ecosystem. Human beings have become oppressive dictators ruling the other creatures of Nature.

The simple way of life that required food just to sustain children and grandchildren has been forgotten. The new life is devoted to making a profit, to fulfilling desire, and to gaining pleasure. An abnormal ecosystem has been created. Selfish desires are becoming ever stronger. It seems people have lost the courage to revert to a normal level of life. They find it difficult to escape from the search for a wealthier, more convenient and what appears within their understanding to be a more pleasant life.

Human beings are born as part of the greater Life of the Universe. The ability to think has been bestowed on us. We are intelligent beings. This requires us to develop our minds, to exercise our excellent ability and technology, to understand the spirit that lives deep inside materials. Material development may well be needed. Nonetheless, developing the mind and heart to grasp the great spirit of the Universe that was bestowed on us is a human responsibility. The Universe contains a fundamental access to the greater Life. We human beings must develop this spiritual potential as well as master the material.

What then is required to develop human spiritual potential? We must grasp the mind of our Omnipotent Parent who has delivered life to living beings. We must learn and know the spontaneous Chain and Cycle of Life, and live wholeheartedly on the flow of past, present, and future, as well as within the current flow of life in the world where our life is given.

Mother Earth has delivered life on the ground. The ecosystem prevails in the air, mountains, rivers, seas, and lands. This blessing through the linkage of all lives deserves our genuine gratitude. Without it we cannot survive. Is there anything worthwhile for us to offer in thanksgiving? Life given to human beings should not remain only with them. Returning the flow is natural. It is a cycle. Life must be returned to the Omnipotent Parent, the Universe.

Instead of expressing gratitude, too many human beings are discharging evil spirits and poisoning the air. Pollution and other harmful impacts on the environment hurt animals and plants. Environmental damage also injures mountains, rivers, oceans, and even the atmosphere. Other living beings are maintaining their lives within the laws of the Nature. Human beings will have to accept responsibility when Earth's self purifying and recovery mechanisms cross the homeostatic plateau.

When we live in communities, we seek to maintain harmony by recognizing the essential requirements of others, and by learning from them. Similarly, we should recognize, help, and embrace the natural environment, and work together to restore a world of natural equilibrium. I believe this is the Will of the Universe, written in the Life Chain and transmitted from the eternal origin of the Universe.

Environmental destruction and social problems caused by current selfish, greedy practices, as we find today, are leading to a path that no longer conforms to the laws of Nature. Human society is at the brink of self destruction. What is progress? What is wealth? Do they really lead us to happiness? How far can we disturb and injure the body of Mother Earth upon whom we and all other beings depend for our very survival? We have been far too arrogant? We have not been thoughtful, responsible human beings.

Our planet is irreplaceable. It is, and ought to be, so beautiful. However, we have caused it to be surrounded by dark clouds of imbalance, by confusion and bitterness. All major problems in the world stem from a human lack of spiritual values; values that should have been promoted in parallel with the material prosperity attained in the history of civilization. As people began to gain material wealth and conveniences, they forgot the fact that human beings cannot survive by contravening the laws of Nature. We cannot destroy Earth that nurtures our life, and survive. When humans' hearts lost contact with the land, people forgot the blessings of Nature and began to overestimate their own abilities. We have deceived ourselves into believing that human beings are the rulers of all. We have come to believe that science and technology, the products of human brains, can reveal and control and develop everything in this world.

In our daily lives, we are surrounded with the products of science and technology. We have come to believe that without these products, social life would no longer function smoothly. While we must give thanks for the contributions of science, we must also recognize that if upside-down thinking about the relationship between Nature and man spreads across human society any further, misfortune will fall on the human community like torrential rain. Indeed, the human community may already have caused environmental damage that is irremediable.

Present environmental issues are urging us seriously to reflect. The red signal is flashing, warning us. We must face honestly and firmly that temporary measures can no longer cure environmental deterioration. We must re-examine our life style. The situation requires clear thinking and firm action. Fundamental measures must be worked out on a global basis. Extinction of the Life Chain would be a profanity for the human race to bequeath to the Universe.

Serious global environmental issues

Today, meetings on environment and human survival convene frequently in many parts of the world. A meeting of scientists, held a couple of years ago in Osaka, predicted "If it goes as it is, the world will collapse in 100 years."[5] This prediction assumed that the world's population, estimated to reach 10 billion in mid-21st century, would enjoy the wealthy life of a developed country. Assumptions included using grains for feeding cattle. Since the projections did not consider freakish weather and other calamities that are likely to occur, the real situation would be further worse. Will we change our life styles in the face of the warnings by scientists? Despite clear evidence that environmental issues are far more serious today, many people continue to act selfishly. True, scientific prowess has also advanced. Nonetheless, in overestimating the abilities of science and technology, too many are not heeding the clear warning. The responsibility cannot be transferred to others.

The term "environmental risk" implies something unknown where cause and effort are difficult to identify. An environmental scientist stated "With respect to the global environment, it is difficult to expect clear explanations from science." This statement is worth keeping in mind. It is often said that science cannot deliver a clear explanation until

---

[5] *Asahi Shimbun*, Tokyo, 14 November 1991 (evening edition).

after everything is over. Then, when the explanation is available, there will be no human beings found on Earth.

Of course, the role to be played by science will become increasingly important. In the field of the environment, in particular, full support from governments will be necessary as research, analysis, technological study, and development woo the expectations of the entire human community. At the same time, people should not overestimate the capabilities of science. They should not claim that scientists alone are responsible for wrong. Ignorance is more harmful.

A scholar once described Japanese economic advances:

> Japan has enlarged its economic size. Like the shell of an egg it has grown so big. However, the thinking of the people, the content of the egg, may not have advanced so far from that in the past.

The statement hinted that a standstill in the thinking of the people is causing various contradictions and friction in the international society.

Using the same analogy, one may say that human beings have made tremendous advances in science and technology; the shell of an egg has become amazingly big. People's thinking, the content of the egg, has not kept pace with the growth of the shell. Thinking is spiritually thinner and more feeble than in the past. Yet it is this spiritually deprived and enfeebled thinking that is recklessly driving the machine of civilization, leading to environmental degradation and devastation of the human heart. What kind of chicks will be delivered to the world from heterogeneous eggs nurtured by human beings?

How can we bequeath freakish, ugly chicks to our children who will live in the coming centuries? Our generation bears a heavy responsibility.

The Moon and mankind

Let's return again to consider the Moon, Earth's satellite. There are some theories on the origin of the Moon. For instance, a parent-child theory suggests that it was made from a part of Earth's mantle. Another, known as the catch theory, posits that it was made at another place within the solar system and caught by Earth's gravity. Yet another, a brother theory, holds that it was made at the same time near Earth's orbit from the same gas and dust that became Earth. None of the theories has definitive evidence, so that the origin of the Moon is not yet certain. Evidence that the Moon's ingredients are similar to those of the mantle of Earth was found by analysis of the moon rocks brought back by the crews of Apollo that were sent up by the United States between 1969-72. Earth's mantle spreads between about 35 km below the ground and about 2,900 km deep, or the above the core. The thickness of the Moon's mantle is estimated to be some 1,200 km, about 40% that of Earth.

In my book published in 1991, *A Message From OISCA To Mother Earth*, chapter 3 "Life fostered by the Moon," I described various functions of the Moon. As we come to know more about the common structure between our planet and its satellite, our feeling of affinity with the Moon becomes stronger.

A project to build a station on the Moon's surface may be launched by 2010. The moon rocks brought back by Apollo contain about 42% oxygen by weight. According to a news report, 200 grams of water were extracted from 10 kg of moon rocks, chemically combining hydrogen and oxygen. This was the result of applying hydrogen gas to the moon rocks. If water can be made, it is possible to produce oxygen. Research to recycle human perspiration and urine to produce water is said to be in progress.

Moon rocks contain 1,000 tonnes of helium $(He)^3$ per 100,000 tonnes of rock. Some 500-600,000 tonnes is estimated to be on the Moon's surface. This volume is the equivalent of ten times the fossil fuels on Earth, such as oil and coal, including those already consumed. Such volume could meet the energy demand of the whole global community at the present level for 1,900 years. Helium used as a fuel is a clean energy source. The potential for using the Moon's helium as a fuel for Earth is huge indeed.

Helium is the second largest element (after hydrogen) available in the Universe. The Sun produces huge volumes of hydrogen by nuclear fusion. Helium is scarce on our planet. Even though the Sun discharges a huge volume of helium, it cannot reach Earth because it is suspended in our atmosphere. Since the Moon has no atmosphere, helium that falls on it accumulates in the rocks and sands.

Space development began in 1957 when the former Soviet Union sent up a satellite called Sputnik No. 1. It led to competition between two super military powers, the United States and the former Soviet Union. Twelve years later, in 1969, the U.S. succeeded in landing humans on the Moon for the first time. The words of Neil Armstrong, an astronaut sent up with Apollo 11 and who landed on the Moon, are still fresh in our minds.

That's one small step for man, one giant leap for mankind.

Twenty years later, on 20 July 1989, at a ceremony to commemorate the 20th anniversary of the Apollo landing on the Moon, President George Bush proposed to begin "the permanent settlement of space." He said

...for the new century, back to the Moon, back to the future, and this time back to stay.

Various tests are being performed at the moment to see how moon rocks can be used as raw materials. Aluminum, iron, silicon, and other materials needed for the construction of a base on the Moon's surface are available in the rocks and sand already there. Using materials already there would be far more convenient than transporting them from Earth.

Space station "Freedom" to be built halfway between Earth and the Moon, will become a base for transmitting solar energy. Space shuttles will support the work with regular service between our planet and the Moon. Space development implies the arrival of an age of space utilization. There are signs that space development is no longer adventure. It is a business with technical possibilities and confidence in their attainment. Japan's success in February 1994 in launching into orbit the H-II rocket with Japanese technology has helped to inspire this possibility.

It is argued that, when fossil fuel energy sources are used up on Earth, it will be too late to explore space development. Action should begin now while we have still time. These arguments carry persuasive power. Without doubt, human beings today are facing many difficult issues as the population explodes, energy resources are depleted, and the environment is damaged. It may be possible with the advanced level of science and technology to use the Moon's resources to cope with some of these current pressing issues.

Space development is a very advanced and progressive programme. Yet, our ardent pursuit of space leads me to question whether human beings have really exhausted their resources on this planet. The Moon's resources may prove to be an attractive way to sustain our human lifestyle. Any suggestion that we should give up exploring ways to use its resources would certainly appear obstructionist. I do not mean to imply that the Moon should remain exclusively a celestial object of romance. However, we should explore other alternatives, including new uses for Earth's resources and the sustainability of our current lifestyle.

We must be continuously aware that the limitations of human vision, particularly the limitations of the materialist world view, have led from one ecological error to another throughout the course of human history. As science and technology have advanced, it looks as if the resources of the Moon's surface may well provide one means of protecting the environment of our own planet. Yet, what will be the actual result of our plundering the resources of the Moon? What effects will it have on Earth? On the Moon? On the relationship between the two? Do we know enough? Is it healthy for the human community to take such a choice? Will future generations praise those of us living in the last decade of the 20th century? Will we have chosen wisely for succeeding generations?

From "Earth Ethics" to "Universe Ethics"

I believe that we human beings who have been given life on Earth by the creator of the Universe are obliged to keep symmetry with the natural environment of Earth. Mother Earth gently provides for human and other beings and all resources through the framework of the Life Chain and Cycle. We should seek to harmonize with the greater life of the

FROM "EARTH ETHICS" TO "UNIVERSE ETHICS"

Universe at large. There is danger in plunging into space development without reflecting on what human beings are doing and have already done on Earth. When approaching space, human beings must have in mind a clear ethic that starts on Earth and extends to the Universe. We must not even consider rushing ahead in its absence.

Today, we speak of a crisis that involves our very survival. I want to urge every reader to think clearly about this crisis and how it came to be. It will then be obvious that survival cannot be based on material elements alone. We must not lose sight of the spiritual element, that, though invisible, supports us. The material and the spiritual are equally important to sustaining human survival. Human beings are challenging the laws of Nature at an unprecedented pace. Those who notice the danger must sound the alarm.

It may be human nature to act precipitously in the name of development when new discoveries are made. If we begin to alter the Moon, however, its function as a "lunar" body may get out-of-order. This could well lead to abnormalities of both body and spirit for human beings on Earth. "Development" can lead life into great danger, rather than towards a sustainable future.

Just as there are "human ethics" for human society, we need "Earth Ethics" to bring all living beings in harmony with Earth. In the Universe, too, it is necessary to observe "Universe Ethics" to sustain the harmonious co-existence of human beings and universal order.

In November 1992, an Asian Pacific regional conference on the International Space Year convened in Tokyo. Eight astronauts including two Japanese conducted a panel discussion with the theme "Earth watched by astronauts." According to a news report, the common impressions expressed by these astronauts was:

We felt that Earth was the mother of human beings. Watching from space, we could not distinguish national borders. We felt that Earth was our native place. Beautiful and rich colors of Earth can hardly be described by camera eyes. Human beings on board the common ship must help each other to live in peace....[6]

I have long been advocating the concept of "Mother Earth". The astronauts' description of Earth as "the mother of human beings" expresses a genuine feeling far beyond logic and quibble. Adding a spiritual element to these genuine feelings would be the basis for "Earth Ethics".

Like the experiences of the astronauts, International Space Year 1992 brought us many lessons along with a number of new discoveries. It helped to focus our thoughts on the Universe, the only place where the terms "universal" and "absolute" are meaningful in their true sense. I firmly believe that human beings must intelligently think in universal and absolute terms for the survival of present and future generations. While we can be proud of the advancement of science and technology, we should be careful not to let our minds become "haughty," lest the fate of other brilliant civilizations of the past awaits us. Now is the time to consider and apply the lessons of history.

Tens of millions of stars like the Sun exist in the Milky Way. The galaxy bulges at its center and spreads about 50,000 light-years on both sides. In the vast Universe, modern science has discovered tens of billions of clusters of galaxies. It is said that there are a tremendous number of galaxies and clusters of galaxies that are far larger than our

---

[6] *Mainichi Shimbun*, Tokyo, 23 November 1992.

galaxy. Nevertheless, no other star or planet has been discovered that has the conditions for life as we know it.

Today, astronomers are trying to find an "unknown Universe" that is believed to exist between the origin of the Universe at 15 thousand million light-years and ten thousand million light-years away. Scientists and astronomers all over the world are competing to be the first person or group to experience an encounter with "the unknown."

The birth of Earth took place even though there was a low probability of such an event. Some people view this event as accidental. Others believe it was ordained. I am one who believes in Earth's inevitability. From the time of the birth of the Universe, the Life Chain has existed and its flow is not accidental.

CHAPTER TWO

Human Life is the Art of God

Human beings as a mini-Universe

The probability of the birth of a human being is very low indeed. Tens of millions of spermatozoon start from the same starting point under the same conditions. Normally, only one reaches the goal where the ovum waits. All the others give up along the way. A severe struggle for survival is fought in the mother's womb before a human being is conceived.

Today's global population exceeds 5.6 billion persons. None has the same genes. Here, we can begin to uncover the mystery, dignity and importance of human life. Biology demonstrates that the nuclei of cells of human beings have 23 pairs of chromosome. The factor that determines whether a child will be male and female is only one pair: the XX type is female and the XY type male. A difference in just one chromosome pair defines sex. How many realize that the determining factor is even smaller—the gene that is in the chromosome?

Nonetheless, "Life" is much greater than that. Human beings, other animals, plants and even single-cell microorganisms all have such genetic structures. Research is planned to search for the substance of Life by deciphering the information carried by the genes. Biotechnology based on the results may cure diseases early and surgically introduce genetic changes. Even the cloning of Life may be possible, though I am apprehensive at the thought.

Scientists are competing for genetic patent rights. We are living in an age when genes—which are not the product of humans—have, nevertheless, become subject to patent laws. Does this conform with the normal flow of the Universe?

Day in and day out, scientists and astronomers maneuver for leadership. Astronomers compete to make new discoveries about the origin of the Universe and Life. Biologists compete for controlling power over genetic research. These struggles cover the extremes of both the macro- and micro-universe.

As long as the competitors are striving to achieve progress for human society, such competitions can be welcomed. Sincere research and the search for truth will flow with the Will of the Universe. Peace and prosperity lie in harmonious advancement within the Life Chain which exists. Selfish motives will need to be set aside by those concerned if they are not to stray from a true course.

Cells are the basic unit of life of all living beings, including microorganisms. Cells are invisible to the naked eye, as small as 10-30 micrometers (one micrometer = 1/1,000 mm). Our body is composed of about 60 trillion cells. Just as a house has a kitchen, bath, toilet, bed room, dining room, drawing room, entrance and other facilities, a cell has all the basic functions needed to sustain life.

A human being can be likened to a "mini-universe." The cells of the human body are like the galaxies and fixed stars of the Universe.

Three functions of cells

Cells have three life functions. The first is to make a copy of itself (children), a self-replicating function of cell division.

The second is to absorb nutrients, transform them into energy and control themselves to maintain a sound condition by means of metabolism. The third function is to take in information from the outside and adjust according to changes in the external environment.

Despite a size so small that the human eye cannot see it, each of the 60 trillion cells that compose our physical body is working to make copies of itself, in the same way that new stars are born in the Universe. New babies inherit their parents' character. We have excellent eye-sight but it is not normal to see a cell function, even if we concentrate intently on our hands or arms. Who then is giving instructions to our cells?

For a fixed star with mass comparable to the Sun, its primitive time—its egg period—is about one million years. The human baby spends about 280 days in the mother's womb. In the scale of the Universe, it is reasonable to think that the birth of a new cell, a tiny being, is but a momentary event. Life seems to repeat itself with similar cycles appearing to be common at different levels—the Universe at the macro, human beings at the medium and cells at the micro-scale. Here, too, we can uncover a bit more of the mystery of the Life Chain and the "Cycle of Life."

Turning to the second function of the cell, that of metabolism and self control, consider a skin cell that goes on renewing itself ceaselessly. As we experience in our daily life, after a minor injury, new skin can grow in a few days. We human beings are animals with a fixed temperature. Our body temperature normally maintains about 36.5° C. When temperature is beyond the normal range, the body's metabolism reacts and can lead to the end of a life. To maintain the required fixed temperature, heat is discharged through perspiration when body temperature rises. In conditions of cold, terminal blood vessels at the skin's surface shrink or quiver and the body moves to generate heat. Our body is made so delicately.

Our human actions are often not as delicate and orderly as our bodies. For example, Japan is the second largest importer of food items, second only to the former Soviet Union. Nevertheless, we Japanese are accustomed to excessive eating. There is so much waste in our eating habits. In Japan, the total volume of rice produced by farmers—working hard with sweat flowing from their foreheads and bodies—is about ten million tonnes a year. Reliable statistics suggest that an equivalent volume of waste comes from our tables. Excessive eating and drinking, imbalance in diet, and irregular life habits result in various kinds of diseases.

At the same time that the depletion of the ozone layer, global warming, and acid rain—to name only a few modern problems—are warnings from Earth, the large volume of garbage resulting from our affluent life style is becoming one of the major polluting agents in Japan. If the current state of affairs of human society is reflected in our micro-sized cells, they must be angrily grinding their own bodies at every point within ours. From the basic functioning of cells, we can learn about our own responsibilities to sustain life. If we actually pay attention to the signs at each level, it is no exaggeration to say that we human beings, living on the eve of the 21st century, are receiving challenging urgent messages from both the macro- and micro-worlds.

What lessons can we learn from the third function of a cell: adapting to the external environment? Advances in transport have made Earth smaller. We can now fly by jet plane to any part of the world. Advancements in information and communication systems enable us, sitting at home, to watch on live TV happenings at remote areas of the Amazon forests, on top of the Himalayan mountains, and at any other part of our planet. It is possible to watch our own planet from outer space—merely a dream only half a century ago. We are living in an age when the shape of the Universe 10,000 million light-years away can be observed through telescopes, revealing the birth of new stars and the death of old ones.

The brilliant advancement of science and technology is exciting. How much and how quickly are human beings adjusting to the changes in their external and internal environments? Consider the population issue. Three out of four people today are living in developing countries. By 2050, it is estimated, nine out of ten persons will be born in developing countries and remain there their whole lifetimes. The great majority of people in the developing world will finish their lives having struggled with poverty, hunger, malnutrition and diseases. The life which matured with an ovum and one sperm, having won the battle of survival among tens of millions, may end without dignity; its important mission as a human being will remain unfulfilled. This thought alone is enough to indicate that human beings have not been able to adjust themselves sufficiently and quickly enough to their changing environment. Serious reflection is due.

Symptoms that emerge when cells inside a human body lose their ability to adjust quickly to the external environment are cancer and other types of diseases. AIDs, Acquired Immune Deficiency Syndrome, is a major focus of the global community. With both cancer and AIDs, only death awaits in the absence of early discovery and treatment. In many cases the symptoms do not appear until too late. On the personal scale, nothing is better than prevention. On the global scale, environmental deterioration is perhaps the most evident symptom of imbalance. Here, too, we must move more quickly. Prevention, early discovery, and treatment are essential.

Poverty, population, and environmental problems

Poverty presents similar issues demanding our attention. Until the beginning of this century, it was thought that population increase would occur only in the wealthy countries and regions. This was believed to be as natural as for water to flow from a higher place to a lower one. Reality, particularly in the second half of this century, defied the experts. Population increase is most evident in the southern hemisphere. The time for effective remedies has already passed. Despite the efforts of The World Bank and other international agencies, as well as concerned countries, there are no positive signs that poverty will be alleviated in the near future.

If the rate of population increase goes on at the current level, it is estimated that the population of Africa, where poverty is already severe, will double in 25 years. Projected population growth for Africa is six times as fast as in Europe, where the population is expected to double only in 140 years.

The outlook is similar for environmental issues. For instance, CFC, chlorofluorocarbon, gas is a general compound in which atoms of chlorine and fluorine have fixed to carbon. When it appeared some years ago, people praised CFC as a fantastic discovery. It was claimed to be non-poisonous, not easy to burn, chemically sound, and corrosion-free. Because of this, CFC became an essential gas, both for industries, and for our everyday life. Its uses as a refrigerant for refrigerators and air-conditioners and as a spraying medium are far reaching and widespread.

In 1974, an American chemist pointed out that CFC discharged to the atmosphere will reach the stratosphere. There, with strong ultraviolet rays, it will dissolve, discharging chlorine atoms. He warned that there was a probability that these chlorine atoms would dissolve the ozone layer as if in a chain reaction. This warning received little attention at the time.

In 1982, Japanese scientists who were undertaking Antarctic research observed the phenomenon of a widening hole in the stratosphere. The "ozone hole" focused new attention on CFC discharges. The scientific community which had failed to heed the warning in 1974 had not anticipated the appearance of an ozone hole in the Antarctic sky. Their attention was arrested like a medicine that, when, taken in moderate quantity provides an excellent remedy, but, when taken excessively, turns out to be a terrible poison, the uncontrolled use of CFC's could not be allowed to continue. There is now an international agreement to control the use of CFC's. A similar agreement exists to restrict the volume of carbon dioxide discharge. Unfortunately, neither agreement is observed by all countries.

Fortunately, our Earth may not be so vulnerable as to completely disappear as a result of human activities; perhaps it will only be human beings who will disappear. Despite widespread evidence and alarm that the global human community may be at the brink of its own annihilation, there is no sign yet that we are willing to accept this "ultimatum" and are willing to take the necessary common steps for our own survival. Alas, we as human beings have yet to come to understand that there is a true Will of the Universe that cannot be violated with impunity.

When we consider the population explosion the following allegory seems relevant:

A lotus plant sprouted in a pond. This multiplied, doubling in a year. One hundred years later, when the lotus plants filled the pond, they all suffocated and died. The question then arises, after how many years would the lotus plants fill half of the pond? The answer is in the 99th year, a year before total annihilation.

Doubling their number every year, lotus plants take 98 years to fill a quarter of the pond surface. Nobody would feel a crisis at this stage. Yet the next year, the 99th year, half of the pond is covered by lotus plants. How many people might still not sense their fate?

I think this story contains a very important lesson for current global inhabitants. Who can deny the possible arrival of a dangerous "99th year" when population increase, global warming, acid rain, destruction of the ozone layer, and many other negative factors are multiplying at the rate they are today? Should it occur, what would be the effect on the "flow of life" which has ceaselessly continued ever since the origin of the Universe about 15,000 million years ago? We must urgently take serious note of the crisis.

The advances of genetic research

As stated earlier, a human body contains about 60 trillion cells. DNA, deoxyribonucleic acid, the essence of genes, lives in the nucleus of the cell. A clear understanding of the DNA structure is a very recent discovery.

DNA is a combination of two strings fixed to each other in spiral form, and joined at the base. When the string of DNA in a cell is unwound and extended, it is said, it can be as long as two meters. Even so, DNA is invisible to the naked human eye. It is contained in the nuclei of cells that are themselves only several microns in diameter.

The structure of DNA surpasses imagination. There are four kinds of bases, with limitless arrangements. Each of the arrangements is known to carry important genetic information. Three billion base pairs are held in the nucleus in symmetry. This is the ultra-micro world. As we come to know that life is governed by such extremely small elements, where strict order and law are observed, we cannot but feel awe and express gratitude for the profound and mysterious power of Nature.

Each cell of a human being contains three billion base pairs. Effective functions have been identified for only five percent of these. The remaining 95 percent are said to not have any genetic effect. Research on DNA base arrangement and function has examined no more than between 2,000 and 5,000 cells.

We must remember that a struggle for patent rights to genes and DNA is going on amongst scientists in advanced countries. Here, we should pay close attention. Further research may determine whether or not that remaining 95% of the genes has an actual genetic role. If DNA is likened to the galaxies and fixed stars, what appears now to be useless may well be like the dark matter that occupies more than 90 percent of space. We may find with DNA that the 95% really supports the whole (allowing the 5% to fulfill its role). Research in this area may hold potential for future advancement. DNA carries genetic information that is the basis of the flow of Life and very important for our existence.

The basic physical structure of animals and plants is made of amino acids. Three thousand five hundred million years ago, amino acids, the origin of life, first emerged on Earth.

With these as a base, such unicellular organisms as primitive bacteria and blue-green algae were born. DNA already existed in these living beings, and it is because of DNA's activities that succeeding generations of bacteria and algae were born and that they exist today.

Surprisingly, there are genes which are common to both *paddy* (rice), the root of Japanese spiritual culture, and human beings. The similarities were discovered by a "Paddy-genom scheme" managed by the Japanese Ministry of Agriculture, Forestry and Fisheries, that undertakes research on paddy genes. Of about 200 effective genes deciphered by the scheme, there are 28 genes common to both rice and human beings. Further advancement of this research may find more common genes.

The existence of common genes indicates that system which perpetuates life within both animals and plants may well have a common *modus operandi*. It indicates to us, too, that the Life Chain, which originated at the very beginning of the Universe, is involved in all our lives and is truly universal.

Birth of life, a sacred event

Anthropology has spotlighted the question, "Did Adam and Eve really exist?" An answer, of course, would determine the forerunners of the human race. One of Japan's oldest classics is the story of *Izanagi-no-mikoto* and *Izanami-no-mikoto*, written in *Kojiki*. The god *Izanagi* and the goddess *Izanami* gave birth to various gods, and to seas, rivers, mountains, plains, plants and grasses. At the end, they gave birth to human beings. To give birth means that there is a blood relationship between the one who gives birth and the child.

Gekko Astronomical Observatory panorama with
Mt. Fuji as background. (Page x.)
(Photographed by the Observatory.)

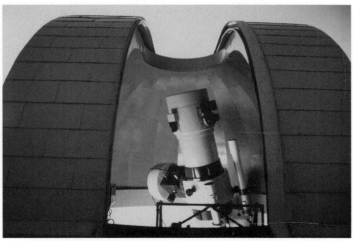

Gekko Observatory's proud 50 centimeter reflecting
telescope plays an active role in discovering asteroids.
(Page x.) (Photographed by the Observatory.)

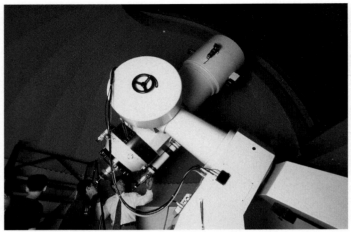

Observatory staff member explains about astronomical observations to visitors. (Page x.)
(Photographed by the Observatory.)

One of the causes for the extinction of dinosaurs some 65 million years ago is thought to be a collision with Earth of a huge meteorite. The photo shows a simulated picture of the shock when a heavenly body of about 800 kilometers diameter collides with Earth. (Page 1.)
(Photograph provided by Reuters-Sun/Kyodo.)

The solar system—centered on the Sun and including Earth—is located in the galactic system (a big group of stars). The diameter of the galactic system is 100,000 light years and the solar system is located 30,000 light years away from its center. The Milky Way (which can be seen from Earth) is the surface of this galactic system. (Page 3.)
(Photographed by the Observatory.)

The Mihozeki Meteorite with a weight of 6.4 kilograms has a history of 4,600 million years. Dr. Masako Shima of the National Scientific Museum concluded that it had hung in space for about 61 million years. (Page 4.) (Photograph provided by Kyodo News Service.)

The Moon is the closest heavenly body for human beings. It has a substantial involvement with human beings and human society, controlling Earth's tides. The diameter of the Moon is 3,476 kilometers or about one-fourth that of Earth. Viewed through a telescope, one can clearly see the craters that have been made by the collisions of meteorites. (Page 5.) (Photographed by the Observatory.)

The Hubble Space Telescope was launched in April 1990. Many new discoveries have been realized with its high precision. (Page 12.) (Photograph provided by Reuters-Sun/Kyodo.)

The Virgo Galactic Group is a huge cluster of galaxies lying nearest to Earth. Some 2,500 galaxies have been observed, of which about 75% are spirals. (Page 12.) (Photographed by the Observatory.)

Photo 1
The launch of ASTRO-D ("ASCA") by M-3S I-7 Rocket
(Feb. 20, 1993, Kagoshima Space Center)

The Institute of Space and Astronautical Science,
Ministry of Education, Science and Culture

The X-ray astronomical satellite launched in February 1993 by the Institute of Space and Astronomical Science of the Ministry of Education has been named "ASTRO-D". (Page 12.) (Photograph provided by the Institute of Space and Astronomical Science of the Ministry of Education, Japan.)

The Milky Way "M81" is located outside of the Great Bear system. Its brightness is 7.9 grade and its distance is 10 million light years from Earth. Since it is an outside Milky Way, "M81" is a big group of stars like our galactic system. (Page 13.) (Photographed by the Observatory.)

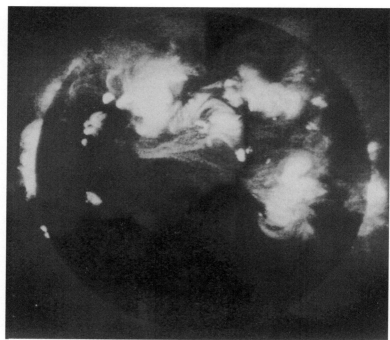

In August 1991, the Institute of Space and Astronomical Science of the Ministry of Education launched a satellite "Yoko" to observe the Sun. Based on over two years continuous observations, discoveries calling for new theories have successively been made, such as that there are two kinds of originating mechanisms for flare and a new mechanism for blowing out the wind of the sun. The photograph, showing a fiercely active sun, was taken through the X-ray telescope of "Yoko". (Page 15.)
(Photograph provided by Kyodo News Service.)

Stromatolites that continue to live by photosynthesis on the shore of Western Australia, said to hold an important clue for investigation regarding the origin of oxygen in the air. Photo shows stromatolites dried up on a shoal. (Page 19.) (Photograph by the Observatory.)

Stromatolites photographed from under the sea. (Page 19.) (Photograph by the Observatory.)

Fossils of psilotinae, discovered in Germany.
(Page 19.) (Photographed by the Observatory.)

Bright future for children. The responsibility of grownup people who live in contemporary times is heavy.
(Page 26.) (Photographed by OISCA-International member.)

A simulated picture of a moon base in its early stages. Seeking for resources on the Moon, human beings may live on the Moon. They will live underground, since space rays, intense ultraviolet rays, and shooting stars fall directly on the surface of the Moon. (Page 27.) (Photograph provided by Kyodo News Service.)

On July 21, 1969, from the Apollo space ship of the U.S., human beings arrived on the Moon for the first time. Captain Neil Armstrong, who was the first to walk on the moon, took this photo. (Page 28.) (Photo provided by Kyodo News Service.)

A simulated picture of the completed internationally manned space base "Freedom" scheduled to conduct space experiments. Although the scope of "Freedom" has been reduced due to the financial situation of the U.S., work on the space station continues to progress. (Page 28.) (Photograph provided by Reuters-Sun/Kyodo.)

Chromosome of a normal male. Since the pair consists of X and Y, it can be distinguished as that of a male. (Page 33.) (Photograph provided by Kyodo.)

The population problem is one of the big issues on the global scale. Photo was taken in Bangladesh. (Page 37.) (Photograph provided by Reuters-Sun/Kyodo.)

Manila, the capital of the Philippines, covered by smog. Pollution issues of big cities have become increasingly serious year by year, particularly in developing countries. (Page 37.) (Photograph provided by Kyodo.)

In the sense of nature that the Japanese people have cultivated, shrine and forests are inseparable. There exist "forests of local deities" nationwide. Photo shows Shibahiko Shrine surrounded by the forests in Miyagi Prefecture. (Page 57.) (Photograph provided by Jinja Shinpo-sha.)

Space without edges. Photo shows the big Orion nebula. (Page 68.) (Photographed by the Observatory.)

The destruction of forests also accelerates the extinction of "species". (Page 83.) (Photograph provided by Kyodo News Service.)

The "Earth Summit" held in June 1992 in Rio de Janeiro, Brazil, was the largest international conference of the century. (Page 87.)
(Photograph provided by Kyodo News Service.)

Participants at a "Sacred Earth Gathering" surround Mr. Maurice F. Strong, Secretary-General of the "Earth Summit", and his wife. The author appears between Mr. and Mrs. Strong. (Page 87.)
(Photograph by OISCA-International member)

Creation is totally a different matter. In Japan, because of such stories and the theory it represents, a way of thinking was born. This holds that land, rivers, mountains, plants and everything that exists in this world are brothers and sisters sharing a common parent who is omnipotent. We live because of Nature. We can live only by means of adapting our behavior to the laws of Nature. This way of thinking, worship, respect and care for Nature have deep roots in the Japanese mind.

Throughout history, the ratio of human beings has been half male and half female. Nobody knows the reason. It is not at all the result of artificial operations. Rather it seems to be the result of Nature's providence. Japanese believe that children are Nature's "gifts from heaven." Life Chain and Cycle are in the hands of Nature. When we follow Nature, genuine prosperity of life will be attained. This is the way we Japanese believe that the Universe works.

Recent developments in medical research are now capable of changing that balance between male and female. The news that artificial selection of male and female has reached the level of practice should give us cause to pause for serious thought. In Japan, at least, in this era of lower numbers of births, female babies have more popularity than males. The current medical technology of artificial insemination seems suited to this preference. Its success ratio appears high for female babies. Controlling the Y-sperm that gives a birth to a male baby is not yet achievable.

When this advance was first announced seven years ago, public reaction was strongly critical. The practice was said to be against the providence of God. As time passed, people became more familiar with it. Resistance is declining of late; and it appears that the number of parents who want to have female babies through this practice is increasing. Although the birth of female babies through this mechanism may not disturb the overall male-female equilibrium, its practice raises serious practical and ethical questions? Can human beings be as successful in maintaining equilibrium as the providence of Nature? Should human beings undertake practices which might well reshape the direction and flow of Life itself? Do we know enough?

As discussed earlier, the self-replication function of the cell is one of the basic activities of life. The cell makes a copy of itself and forms succeeding generations. There are men and women in our society who are suffering from sterility. Couples who badly want children are trying to have them by artificial means: in vitro fertilization (IVF), surrogate mothers, third party sperm, and artificial insemination devices (AID). At the other extreme we have abortion and contraceptives: the means not to have babies. I worry that the advancement of medical science and the development of treatment technologies are leading to excessive confidence that the birth of new life can be controlled by human beings.

In early 1993, a Japanese newspaper reported that a Japanese female student studying in the United States registered as a donor of ovum with a private infertility organization.[7] The number of babies born of a donated egg inseminated in vitro by the husband's sperm and surgically introduced into the wife's uterus had reached 2,000 in the U.S. alone. While such a baby is genetically the child of the husband and the egg donor, the wife can become a "real" mother who physically gives birth to the baby.

Although Japanese scientific society does not recognize such treatment for sterility, where demand exists, supply usually

---

[7] *Yomiuri Shimbun*, Tokyo, 26 February 1993, evening edition.

follows. The U.S. sterility center alone is reported to hold in ready supply eggs from several hundred donors. When we learn that the cost per egg is $2,500, we must recognize that the danger exists that the humanitarian words are only a mask for the fees, that are the real incentive.

For couples who have not been gifted with children, despite their earnest desires, it may sound insensitive to say that our ancestors, who lived at times when neither in vitro fertilization nor artificial insemination skills were developed, faithfully accepted life as God-given. Nevertheless, the natural cycle of the Life Chain is real and must not be ignored.

Determining the sex of a child raises a different set of questions. It has nothing to do with sterility. It is not that these couples cannot have babies. Rather—in Japan—these couples wish to have daughters, not sons. Of course, it can validly be argued that sex determination practices may be appropriate to avoid certain inherited diseases. Both fertilization and sex determination practices are designed to give birth to "new life" by artificial means. Whether "God's path" and the "human path" can co-exist is a very difficult issue.

Human beings are organisms for whom it is natural to have birth as well as a death. Death is often described as "eternal sleep" or "pulling away the breath." Three symptoms have generally been recognized as human death, namely: stoppage of the heart; stoppage of the breath; and opening of the pupils. Death can be determined rather easily, even in the absence of knowledge of medical science. However, the advancement of medical science and biotechnology is moving to change our traditional view of death. By recognizing brain death, transplanting of fresh organs becomes possible, and through it, the life of another person can be saved. Ordinary people cannot judge brain death; the heart is still beating and the body is warm. Nevertheless, a judgment of death is made by a doctor who anticipates that, since the brain is in a certain condition, the heart, lungs, and eyes will not resume normal operations. If we take an extreme view, the fate of man now lies in the hands of man. Man has wrested fate from the hands of Nature.

Notwithstanding, there are other implications. If brain death is recognized by law and accepted in society, the marketing of organs as commercial commodities will emerge, just as sperm and ovum banks have proliferated. Then, greedy and wealthy persons will be able to buy "new life" in the form of younger and healthier internal organs and the poor may be placed in the position where they need to sell their organs for the survival of their family. Or worse, the poor may sell their children's organs for their own comfort or survival.

Related issues should be kept in mind. The birth and death of human beings are related to the breath of Nature, so that they flow and cycle naturally. Just as a death is described as "pulling out the breath," the ancient Japanese concept of death was that the person was taken back to the hands of the supreme parent. Most Japanese continued to believe in the existence of a world after death. There are traditional customs in Japan to pay respect to the souls of ancestors. Therefore, it is natural for Japanese people, at least, to be hesitant and reluctant to recognize brain death. The questions arise, can "God's path" and "human's path" co-exist and will Nature forgive man's challenges? In dealing with the subject of brain death these are very difficult issues.

My own conclusion is that life should not be controlled by artificial means. The right path for human beings is to accept the Life of the Universe and live according to the flow of

Nature. *Hito* (man) is bestowed the intelligence to recognize and accept the invisible spirit.

Research must be "humanly" motivated

It is said that DNA research will bring about medical cures for inherited diseases by detecting them at the time the fetus is small. Such cures are predicted when DNA's function may be deciphered more fully. A grand plan to analyze the functions of all the genes of human beings, addressing cures for inherited diseases as one of its objectives, is known as the "human genom scheme." As analysis of human genes advances, it will be possible, say experts, to transplant the genes of a healthy person into the body of a person diseased due to abnormal genes.

As is clear from observing natural phenomena, even if we feel we have sufficient understanding, it is always possible that entirely different results than we envisage may occur. We may think that we can cure all diseases as the secrets of the genes may become fully discovered. But do we know what new diseases may spring up in their place?

As long as gene diagnosis and cure go side by side, it is possible that problems may not arise. However, should the technology to cure disease lag behind diagnosis and the generation of more genetic information, problems are likely. AIDs patients often suffer discrimination and are denied treatment as a result of fear and misunderstanding. Even if genetic cures are achievable, how many patients will be able to get their benefits? Will the mass of underprivileged be content to see the wealthier class get priority? What social problems are we creating?

Given the present spiritual development of human beings, social and ethical problems are likely to emerge endlessly. Any number of issues, such as protection for the genetic information about each individual, will trigger terrible problems. Wisdom and spiritual development will be very much needed. To achieve this we must address our education programmes.

Today, Japanese enjoy the world's greatest longevity. Statistics show that by 1997, the population over 65 will surpass those 15 years old and younger. Another projection envisions a super-aged society only 30 years later, in which the aged population will represent one out of every 3.67 persons. Gene research claims that "aging" can be overcome by treatment. If such external factors as the environment and diet are improved, 200 and 300 year life spans, they say, may not be impossible. For me, extension of the life span of human beings through artificial means sounds as foolish as determining the sex of babies. If Nature has not seen fit to provide it, should we presume to know better? Should we not be satisfied with medical care for sickness alone?

There is another critical danger in genetic treatment. If a certain gene is transplanted, the person's reproductive system will pass it to the following generations. Introduced genes may spread to all human beings. This opens the way not only to ideas of an ideologically new mankind, but also to a new mankind with altered physical structures. Is it not dangerous to think of restructuring human beings? Is this not a clear case of the arrogance of ignorance? Such conceit profanes the Will of the Universe.

Nevertheless, I wish to express my profound appreciation to those who have enthusiastically been trying to discover the mystery of life. Knowledge of it does not mean we need to tamper with it. As the term "fate" indicates, life is God-given.

So, humility in life is necessary. Any reckless "hot-rod" action can result in irreparable circumstances.

If damage occurs from genetic tampering through the conceit of human beings, the Life Chain, ceaselessly operating ever since the origin of the Universe fifteen billion years ago, may be severed. Human beings may not be annihilated by their such actions, but they may become non-human beings. The Universe before the big bang is often called "God's sphere." I appeal to the researchers of our world to exercise moderation and consciousness in their efforts to decipher genes. Do not presume to sufficient knowledge of the root of life. Trust in the superior knowledge of "God."

Activities of the ultra-microworld

Genetics is only one aspect of our micro-world. There is another academic subject dealing with the micro-world. Research into the creation of the Universe—the extreme macro-world—starts with this micro-world study of "quantum physics." The term "quantum" is seldom used in our everyday life. In dealing with atoms and molecules, there is a unit known as the "quantum". By way of example, when population is counted, fractions such as 0.5 person or 1.5 persons do not arise. So, in the case of human beings, one person is the minimum unit, a quantum.

While air and water are essential factors for our life, light, too, is equally needed for survival. We see the world with the help of light. Light is a source of thermal energy. The initial temperature of Earth is said to be minus 18° C. If sun light had not been available, Earth would have become an ice planet where no human beings nor most other beings could survive. Because of the light from the Sun, the average surface temperature of our planet is sustained at 15° C, and human beings and all other organisms are able to live.

Quantum mechanics instructs us that light is both "wave" and "particle" at the same time. It is not at all difficult for us common people to understand that light is a wave. Because of the existence of radio waves, we listen to radios and watch TVs. It is easy to think of light as an extension of radio waves and similarly invisible. Ultraviolet rays are also waves of sun light. Most are suspended by the ozone layer, so that humans are shielded from the harm they can effect.

Today, microwave ovens are widely used in homes. With them, fish can be baked without causing smoke. Rice wine, *sake*, can be warmed up. Infrared and ultra-infrared radiation cannot be seen by the naked human eye. Fish and other cooking items are baked or boiled as light waves are converted to thermal energy. So, light is a wave.

Understanding light as a particle is not so easy. Albert Einstein received the Nobel prize for his work with light. He named the light particle the "photon". So, light is both a particle and a wave, Einstein concluded. In the super-microworld of quantum mechanics, it is said, both particles and waves co-exist. One cannot look only at one of them.

The new cosmology emerging from quantum physics

Phenomena of the super-micro world have played a key role in the discoveries regarding the creation of the Universe. Since matter is both particle and wave in quantum physics, it can pass through a very thin partition. The passage of a particle through a potential barrier, even though it does not have enough energy according to classical physics, is a phenomenon called "tunnel effect." At the time of the big bang, a seed of the Universe passed through a partition that was covering the seed, by means of tunnel effect, and the Universe began to expand. This is what scientists imply. In our everyday world, this kind of tunnel effect cannot happen. For instance, a seed of grain cannot come out while the hull remains, no matter how small the seed.

Scientists suggest that the Universe at its origin was smaller than the seed of grain that we see in our daily life. So, if quantum mechanics had not emerged, information about the creation of the Universe, much less the origin of life, might not have been discovered to this day. Quantum mechanics and the theory of relativity are recognized as the most important driving forces for physics in the 20th century.

An important impact of quantum mechanics on cosmology is the discovery of fluctuations in emptiness. Classical physics held that energy is zero in a void. Electromagnetic waves were believed not to exist. Nothing would exist under such conditions.

Quantum physics suggests that fluctuations exist even where there is emptiness. Was there only emptiness? Did nothing exist at the time of the birth of the Universe fifteen thousand million years ago? Quantum physics suggests the possibility that there were fluctuations in the seed of the Universe.

What mysterious fate causes the study of phenomena of the super-micro world to give life to subjects at the extreme opposite? In this vast Universe, the Life Chain is ceaselessly and endlessly continuing. This is evolution itself. How wonderful it is to be living in an age when such mysteries are in the process of being scientifically proven.

The limitlessly expanding Universe

How long will the expansion of the Universe that has been going on for 15 billion years continue? On one hand, there is a theory that the Universe, which started with a big bang singularity, may begin at some time to contract—leading to collapse in a "big crunch." On the other hand, there is a body of evidence that suggests that the Universe will continue to expand eternally. Data received from the observations of the Hubble space telescope, rotating in outer space, proves the latter. The conclusion that the Universe would continue to expand was announced in 1991. In early 1993, after observing radio waves at the end of the Universe, the National Radiowave Astronomical Observatory, USA, announced that the expansion of the Universe will continue, albeit at a slower rate. Contraction will be avoided before it occurs.

The origin of the human race is placed about three million years ago. Compared with the life of the Universe, the length of human existence is nothing but a "dot." Even were it to come to an end after billions of years, the Universe is as if infinite, virtually without beginning and without end.

Earth shares a common fate with the Sun

The Sun's diameter is 109 times the diameter of Earth. The distance between the Sun and our planet is about 150 million km. Sun light reaches Earth in about eight minutes. These are facts known to almost everyone. The Sun is a big ball of hydrogen gas and ceaselessly nuclear fusion reactions take place in its center, producing energy, by which blessing our life on Earth is sustained. A huge volume of hydrogen is burning. At the center of the Sun, the burn rate is 700 million tonnes/second. It is said that the Sun is losing mass at the rate of five million tonnes/second. Even then, its weight has reduced only by one four-thousandth during the past 4,600 million years. In about five billion years from now, the hydrogen in the Sun will have burned out. Then, it is forecast, the Sun will expand and gases will flow away from its surface. Finally, the Sun will become a white dwarf. Notwithstanding, can we not say that the Sun is eternal when viewed from our scale?

The fate of our planet is linked with the Sun. When the Sun comes to its final stages, it is believed that its surface will expand close to Mars. Earth, closer to the Sun than Mars, will surely be absorbed by the expanded Sun. Since such a scene is not anticipated for five billion years, life on our planet, too, can be regarded as eternal.

Connection of spirit transcending space and time

In comparison with the colossal scale of the Universe, Earth is like one of the "cells" of which our physical body is constructed. Each cell is playing its role, just as each planet plays its role. The role of the cell "Earth" is remarkably important. In the human body, diseases like cancer and others emerge when a cell loses its function. Unless there is an early cure, pathogenic bacteria-carrying cells will spread throughout the body and, possibly, a human life will end. As we look around our planet, various cell-eating symptoms, such as environmental degradation, racial fighting and poverty, are evident. When the Earth-cell loses its ability to cure itself, what kinds of ill-effect will Earth bring to the body of the whole Universe?

In the Earth-cell, each human being can be compared to "DNA". The gene can be compared to the "quantum" of quantum mechanics. Remember that quantum has the dual characteristics of "particle" and "wave". Consider the "particle" as our "body" and the "wave" as our "spirit". The Universe was created by its wave passing through a wall. This now well-known tunnel effect, because of spirit, has a connection with the very far past and future, transcending space and time. This connection of spirit transcends space and time. Perpetuating the connection is the spiritual foundation sustaining the cycle of our life. The being who can be aware of this connection is truly a "human" being who is never alone in this never-ending vast Universe.

Return to Nature's law and order

Cells, genes (DNA), molecules and atoms exist in strict order with fixed mechanisms that have been scientifically proven. To date, no human force has been able to influence the formation of such order. Human beings, too, are born within the strict order of the Universe—an order, as I have said previously, that is not governed by chance. The same applies to all other living beings—more than ten million species on Earth. The Life Chain, is part of this order. Any being who would disturb this order will face the natural consequences, that, also, exist within Nature's order. It seems that human beings alone—we who have been given intelligence and consciousness—are ignorant of the Will of the Universe. With escalating conceit and desire, human beings are becoming wanton violators of Nature's law and order.

At its very beginning, the Universe pulled a curtain to launch the great drama of Life. Modern science is seeking to discover the seeds of that birth. There are some who would argue that the beginning of the Universe is also the origin of the flow of Life. That may be.

RETURN TO NATURE'S LAW AND ORDER

Whatever science may be able to tell us, we as human beings must accept with humility that somehow, somewhere, something took nothing and made something. That power, that understanding, is beyond us. Science is extending its reach to the macro-world as well as to the micro-world. As investigations advance, ethical issues will emerge. The cliché "too much is as bad as too little" suggests returning to the path that the law of Nature advocates. While we should be proud of the brilliant attainments of our species, at the same time we should recognize our own limitations and that there is a path from which we stray at our peril.

We are now living in an age of science where the mysteries of the Universe are probed using high-tech equipment. The lively activities of evolution and progress that we observe have been endlessly going on in the large scale structure of the Universe. What were the feelings of our predecessors for whom neither high-tech equipment nor telescopes existed to watch the sky?

In Japan, our forebears in the remote past lived in harmony with the forests. Without the tools and equipment of our modern civilization, nonetheless their gaze was upward. Our classic documents relate descriptions of the high, wide, and endlessly expansive sky and constellation. The Universe was *Taka-amahara,* literally, a plain of high heaven, a lofty and sacred world where the deities live. *Taka-amahara* is the finest expression of admiration and gratitude to the sky where Sun, Moon, and numerous stars shine.

Our ancestors' description of the sky was not just as space, but the place where *kami tsumari masu*—a number of deities live. Using the word *kami,* deities, they admired the great Universe, so full of limitless force. They expressed admiration for the beauty and dignity of the Universe, using *nori goto,* or *norito goto,* ritual words addressed to the omnipotent beings.

Today, there are formal Shinto prayers, *norito,* addressed to the omnipotent.

It is pity that in an age when science has made such remarkable advancement, the modern inhabitants of Earth value materialistic thinking above the spiritual. Except for a limited number of experts, the majority of inhabitants do not watch the sky. Their views of the Universe and Nature are superficial. Most people are becoming selfish. Their thinking is inward. Though they were without scientific knowledge and modern technology, our predecessors in the remote past had a far more broad-minded view of the Universe. They grasped the true meaning of Earth in their life and respected Nature. As we think of our ancestors, we find a lesson. What do advancement and progress mean to human society?

To my great pleasure, the universal perspectives described by our forebears in the remote past are being confirmed as truth by the advances of space technology. There are scientists who say that science in the western world is at last reaching close to "Genesis." The essence of "Genesis," the beginning of the Old Testament, is about to be revealed through the advancement of space science and technology.

In the coming Chapter, I will describe the roots of Japanese thinking that gave birth to a sacred expression, "*Taka-amahara ni kami tsumari masu*" (A number of deities assembling in the lofty sacred Heaven).

CHAPTER THREE

Roots of the Japanese Outlook

Culture closely linked with the forests

The forebears of our Japanese race lived in a mountainous archipelago and maintained a simple life surrounded by forest. Their life revolved around hunting which provided the necessities for life. They recognized that harmony within Nature sustained the flow of the food and life chain. Food, clothing and shelter were blessings provided by the forest and delivered in accordance with the four seasons. Their life styles and culture were closely associated with the forests. Their life was, as I imagine, happily blessed by a gentle climate, never too hot nor too cold.

As long as people lived in harmony with Nature, blessed by their natural surroundings—mountains, rivers, seas and sky—selfish and haughty minds didn't develop. The old saying, "mountains and rivers act violently when they get angry", implies that Nature is an awesome being. Our ancestors found ways to protect their lives by exercising the means available to men to calm and to care for it. Current forest and river conservation programmes are based on the spirit inherited from our forefathers which requires that we progenitors pay respect to and live with Nature. A sense of worship was an integral part of our ancestors. They did not choose to do things for their own human convenience. Rather, they protected Life by respectfully worshipping the deity or deities that they believed existed in Nature.

This tradition of respecting and worshipping Nature in daily life paved the way for "*shinto*" as it is known today. When Continental cultures entered Japan some 1500 to 1600 years ago, different approaches were introduced to Japanese for the first time. The indigenous faith was named "*shinto,*" the way of *kami* (deity) or "divine path". There is no founder, unified doctrine or sacred scripture with *shinto*. For the ancestral people who lived a simple life in harmony and agreement with Nature, the term "*shinto*" never existed. While it is true that Japanese culture has made tremendous progress through the introduction of learning, knowledge, and skills from the Continent, it is also necessary to note that the basic infrastructure for cultural assimilation pre-existed their introduction.

"*Shinto*" is also called "*kannagara no michi*", meaning a human life or a way of acting in accordance with the will of the deity. The name implies an awareness of the divine, the solemn, and the sublime. It carries with it an obedient, sincere attitude towards life in accordance with the laws of Nature. In addition, "*kannagara no michi*" carries awe, respect, and gratefulness towards the soil, water, mountains, rivers, seas, skies, Sun, Moon, and stars. Each was seen to possess an extraordinary virtue essential to human beings and all other living beings, as well as to the life of the Universe which they believed and we believe exists behind it all. It also involved admiration and yearning for their beauty and gentleness. All these factors have been passed down through the generations to today.

Views on Nature nurtured by Japanese

Although *shinto* is considered as one of the world's religions, it is not a religion in the true sense of the word. *Shinto* is a faith that emerged naturally from ancient times when people regarded Nature with complete respect. While Confucianism, Buddhism, Christianity, and Islam, to name a few, are known as ritual religions, *shinto* may be called Nature's religion. With no founder nor doctrine since its emergence, *shinto* has become the focus of the culture and spiritual life of the Japanese people. Ritual religions have founders, who guide people and whose messages become dogma.

The lives of rural Japanese are rooted in their primitive faith. This is not unique to Japan. Apart from the established religions, a deep but quiet worship of Nature guides the life of rural people throughout the world. The Aborigines, for instance, are a good example of a people whose culture teaches them respect for Nature.

In ancient Japan, mountains or hill-tops were sacred places where the forebears of our ancestors were believed to be sleeping with the deities. So, with the genuine belief that life would return to the sacred places, fearful as well as loving sentiments towards the mountains and hill-tops were deeply ingrained in the hearts of our predecessors. As time passed, shrines were built to pay respect to the deities. Most shrines were built at the foot of small mountains or hills, surrounded by thick forests. There was always water flowing nearby, giving the comforting murmuring sound through which the Japanese are intimately linked to the rhythm of Nature throughout the changing seasons.

Festivals and ceremonies based on these earlier understandings remain with us today as the foundations of our Japanese culture and tradition. Only the Emperor surpasses the ceremony organizers in respect.

What I am trying to describe here is the inheritance of deep spiritual values and regard for all of Nature that is inherently part of all Japanese regardless of their outward behavior.

Stephen Hawking says of our time

> Up to now, most scientists have been too occupied with the development of new theories that describe *what* the Universe is to ask the question *why*. On the other hand, the people whose business it is to ask *why*, the philosophers, have not been able to keep up with the advance of scientific theories.[8]

He is exactly right. Philosophers question "why does the Universe exist?...why do human beings exist?" While philosophers may spurn the advances of science and scientists may scorn the methods of the philosophers, surely answers to these questions would temper the more runaway tendencies of science.

I am impressed by Hawking's confidence that science can discover the beginning of the Universe.

> The idea that space and time may form a closed surface without boundary...has profound implications for the role of God in the affairs of the Universe. With the success of scientific theories in describing events, most people have come to believe that God allows the Universe to evolve according to a set of laws and does not intervene in the Universe to break these laws. However, the laws do not tell us what the Universe should have looked like when it started—it would still be up to

---

[8] Hawking, Stephen W., *A Brief History of Time: From the Big Bang to Black Holes*, Bantam Books (New York:1988), p. 174.

God to wind up the clockwork and choose how to start it off. So long as the Universe had a beginning, we could suppose it had a creator. But if the Universe is really completely self-contained, having no boundary or edge, it would have neither beginning nor end: it would simply be. What place, then, for a creator?[9]

While he questions the presence of a creator, Hawking suggests that the mind of God can be understood by the "victory of reason".

> However, if we do discover a complete theory, it should in time be understandable in broad principle by everyone, not just a few scientists. Then we shall all, philosophers, scientists, and just ordinary people, be able to take part in the discussion of the question of why it is that we and the Universe exist. If we find the answer to that, it would be the ultimate triumph of human reason—for then we would know the mind of God.[10]

The paradox seems unnecessary. While questioning whether there is a place for God, he is undeniably conscious of God. Hawking is a Christian, a believer of a monotheistic religion, and the spirit of the Creator may well be deep in his mind. This questioning often seems as if it were a struggle between God and academia. Yet, if Hawking can posit a Universe with neither beginning nor end, what better argument for God, who is without beginning or end?

Interestingly, the word "God" seldom appears in science journals or books written by Japanese. In the thinking of

---

[9] Hawking, Stephen W., *A Brief History of Time: From the Big Bang to Black Holes*, Bantam Books (New York:1988), p. 140.

[10] *Ibid*, p. 175.

most Japanese, this does not mean that God does not exist. Many Japanese may well claim that they don't belong to any religion and are not interested in the existence of God. But, twice a year, they involve themselves in spiritual practices. Just after midnight on December 31 when the new year arrives, more than 80 percent of the people throughout the country visit shrines and temples with their families to pray for good fortune for the coming year. In August, the season for "*obon*," a traditional Buddhist practice, arrives. Families go back to their ancestral homes in the countryside to visit the graves of their ancestors and to pray for them.

From what spiritual or mental motivation do these customs continue? It is difficult to know whether such a high proportion of the population is only following custom, whether it is a Japanese habit, or whether there is some other motivating force. Whatever the reasons for this behavior may be, metropolitan Tokyo becomes quiet on these two major occasions while many residents go back to their native places. Emissions from cars decrease and the beautiful blue sky reappears—a very rare atmospheric event in Japan's capital. Regardless of the substance of worship, it is a beautiful aspect of Japanese culture to stand side-by-side, holding hands, in front of God or statues of Buddha at shrines or temples and at the gravesides of our progenitors.

Of course, such practices are not limited to the Japanese alone. The people of many nations and races in all countries have similar customs. Whatever their motivations or religious faith may be, simplicity and the beauty of worship are common to all. Even a person who has committed the worst crime may not be punished at the moment he worships obediently. A scene of worship is beautiful. The flow of Life and the invisible and visible worlds meet at that moment. Worshippers are obviously in a spiritual attitude. Their thoughts are not focused on material concerns.

Non-Japanese people are sometimes puzzled by the behavior of Japanese who profess to be non-religious and without faith. Japanese youngsters, for instance, often are deeply affected by the indigenous culture and lively practices of people in developing countries, particularly in rural areas that they may visit. The inherited practices of human beings transcend borders. Sleeping consciousness in the minds of Japanese youngsters may be awakened and harnessed by these encounters. On the other hand, modern Japanese are drawn into the speedy flow of civilized society. Everybody looks so busy. Yet, elements unconsciously planted in our genes awaken twice a year and encourage us to return to practices that may seem somewhat strange to foreigners. Perhaps if they could look back to the remote past they might better understand the roots of Japanese thinking and thereby better understand what currently appears to be fuzzy thinking concerning our attitudes toward religion and faith.

The impact of modern thought

Japanese often give the impression to foreigners of being difficult to understand, of being "fuzzy thinkers," ambivalent, and unable to say no. As a result, the Japanese may be put at a disadvantage. While they are excellently adaptable and harmonious with Nature, they are less adaptable to international debate. Traditionally, Japanese society has encouraged less talk and more work.

Yet, traditional values are receding. A number of issues are forcing Japan into the arena of international debate and cooperation. Education must now play a particularly important role. A positive policy to raise Japanese consciousness about global matters is urgently needed.

It has been suggested that the "vague" or "fuzzy" attitude of most Japanese stems from their adapting to and harmonizing with nature. This is doubtful. It is more likely to be the result of modern thought colliding with the thinking practices inherited, unconsciously, through generations. Such collisions can produce disorientation or "fluctuations" and lead to diminished self-confidence. Their influences then become negative. This is an important issue for Japan today. We must find ways to put the impact of modern thought on inherited traditions to positive use.

On the one hand, the conveniences of civilization are benefiting families and making life comfortable. On the other, the human equivalent to desertification is spreading through our society. Misunderstanding of the importance of each individual is causing estrangement among people. The moral code that should bind society and make it beautiful and pleasant is declining. Public spiritedness, respect, humanitarianism and friendship can hardly grow when people live as they please, without paying attention to the needs and rights of others. Our current affluence clearly exceeds that of the past. Nevertheless, people are feeling uneasy and loneliness and mental illness are on the increase.

The more affluent Japanese seem to be ignorant of and show little concern for the wider global society. They have little knowledge of their own country and are indifferent to others. An empty mind does not inspire regard for Nature or reverence for Life. Instead, spiritual values decline. In that sense, they are non-religious and non-faithful. If a man becomes a bank of material knowledge alone, he becomes a poor-minded, cool-hearted being. What, then, can be his *raison d'être*? Why was Life given through winning the spermatozoon race to survive?

Life's roots do have a spiritual or blood relation with the greater Life of the Universe. Spiritual elements can and do communicate and respond with each other. Man's *raison d'être* is that he holds the spiritual element. *Hito*, a man, can be interpreted to mean a being who retains spirit or soul. The Japanese classic "*Nihonshoki*" speaks of "*rei*" as *hi* of *hito*. *Rei* implies spirit or soul, both invisible; and *to* of *hito* means to retain, *rei*. As we search for the roots of the Japanese culture, the *shinto* perspective can offer us many insights.

It is worthwhile to study the views of Life, the World and the Universe that we have inherited. Immediate past generations have accepted culture, learning and thoughts from the West. They took the better aspects on offer and discarded some of the ones inherited from their own ancestors. Nevertheless, the elements that could not be discarded have completely survived in our Japanese spirit, structure, life, and customs. Inadequate digestion, an improper equilibrium between introduced and indigenous ideas and culture may well lie at the root of current Japanese fuzzy thinking.

Kojiki informs about ancient Japanese thinking and life

"*Kojiki*" and "*Nihonshoki*" are two leading Japanese classics. They were compiled in the early eighth century. Looked at from the point of view of the wider context of history, the eighth century is not so far in the past. "*Kojiki*" has a special value as a record of traditions from the far past. It is a kind of dictionary to inform us about the life and thinking of earlier times.

In a cave in Nagasaki Prefecture, archaeologists found the oldest earthenware so far discovered in Japan. It dates from about 12,000 years ago at the time of the *Jomon* period. Tools, older than the earthenware, believed to be from the

Stone Age about 30,000 years ago, have been found in many parts of the country from Hokkaido to Kyushu island. However, a discovery in May 1993 surprised many scholars. This comprised 45 stone tools, excavated at Takamori in Miyagi Prefecture, believed to be from 430-610 thousand years ago. They are of the same era as Sinanthropus. The age of the layers of soil from which the stone tools were excavated has been measured scientifically by three different testing means. Excavations at many parts of the globe are giving us information about the past. Scientific advances are bringing three or four million year old human history into our present age. Ancient and modern are co-existing in human consciousness, as if they are proving that Life is linked ceaselessly through the ages. If we can learn from this to take a balanced sustainable path into the future, such academic research will indeed be useful to the human community.

Returning to the main thrust of my earlier commentary, "*Kojiki*" begins with the phrase, "At the beginning of *ametsuchi* (literally, Heaven and Earth, meaning the Universe), *taka-amahara ni narimaseru kami no mina* (the name of the deity who dwelled in the plain of high heaven) is *Ame-no mi-naka-nushi no kami*, (the deity in the center of Heaven or the supreme omnipotent deity.) Next came *Taka-mi musu-hi no kami* and then *Kami- musu-hi no kami*. The three deities, with *Ame-no Mi-naka-nushi* as the supreme deity, are independent and invisible. It further writes that *Izana-gi no mikoto* and *Izana-mi no mikoto* gave birth to the land, to other deities, and to mountains, rivers, and plants. It describes that *Izana-gi* (the male-deity) and *Izana-mi* (the female-deity) are our ancestral deities.

Whether or not one likes or agrees with these descriptions, they are part of the authoritative classics which define Japanese characteristics. Certainly, there will be some who would say that they had never thought of such things. That may well be the case, but these factors have already become part of the flesh and blood of the Japanese people. Without being in their consciousness, they have been passed on from parent to child, from child to grandchild, and so on. They are alive in the traditions, culture, customs, and even in our daily life.

Naru of *taka-amahara ni nari maseru* (*naru* is the infinitive of *nari*, a verb) has the same meaning as "become," as a child becomes an adult. Just like a baby that continues to grow after birth until finally becoming an adult, *naru* means ceaseless growth and advancement, day-in and day-out. It also means the endless growth of new life.

Modern science has concluded that the basis of all materials and energy existing in the Universe was made by the big bang about fifteen billion years ago. Our Universe is expanding. Soon after its birth, the Universe spread in all directions at unbelievably high speed. This process created elementary particles, electrons and atomic nuclei. Hundreds of thousands of years after the big bang, the first atom, hydrogen, was born. Hydrogen changes to heavier elements by means of burning inside the stars. A variety of additional elements are born by the explosions of stars that come to the end of their lives, called supernovae. The Universe is the root of growth and advancement of all. In the Universe, stars are born and die, and through repeated processes produce complex elements one after the other.

Take the example of planets in our solar system that are rotating and revolving in space. If each of them was to move around freely, they might collide with one another. Since they are moving around in an orderly way, with the Sun as the immovable center or axis, they don't collide. Of course, some small stars or planets, such as singular meteorites, are not so orderly and do collide with other bodies in space.

It is now possible to determine scientifically that movement in space due to gravity. But people who lived in ages when the scientific knowledge of astronomy did not exist were nevertheless still able to recognize the existence of orderly activities in the Universe. Truly impressed by the order of heavenly activities, they said "*Ame-no 'mi-naka-nushi' no kami.*" The supreme deity of the center of Heaven governs the activities of the Universe. (*Ame* means Heaven, *minaka*, the center, *nushi* the head, and *kami* is deity.)

Musuhi, as in "*Musuhi no kami*," literally means germinating spirit. Our ancestors offered the name *Musuhi no kami* to a deity whose function was the bestowing spirit to every living creature.

A phenomenon or event occurring beyond the ken of human intellect is described as *reimyo* or *kushibi*, literally meaning superhuman or mystical. For instance, one meets another without any notice or expectation and both become close friends as if they had known each other for years. Such wondrous encounter is often described as strange or unusual, a relationship that stems from the superhuman.

From the foregoing, *musubi* (with the same meaning as *musuhi*) can be understood as *Kushibi*, activities designed to let everything grow beyond human direction and ability. The germination of moss (*koke*) and grass (*kusa*) is described as *koke musu* and *kusa musu* in Japanese. *Musu* describes a circumstance where something naturally emerges out of nothing and gradually grows up. The function of fruit bearing is called *mi* (fruits) *ga musubu* and the appearance of dew, *tsuyu* (dew) *ga musubu*. The function of growth and progress through natural means is described as *musuhi* or *musubi*. *Musu* and *musubu* are infinitives of *musuhi* and *musubi*. It is not without meaning that, in Japan, a son is called *musu-ko* and a girl *musu-me*.

Musubi also carries the meaning to tie together more than two things into a united form. Tying a string, establishing a friendship and arranging a relationship are called *musubu*, a verb form of *musubi*. To *musubu* is to unite more than two matters and affairs, the result of which is the birth of something new. This is the *musu* function. In Japanese traditional custom, a rice ball is called *o-musubi*, that is made by two holding hands. *Ine* (a rice plant) is phonetically in Japanese, *inochi no ne*, the root of life. Rice sustains our life eternally. When we form a ball of rice and eat it, full of spirit (*ine*), with gratefulness to Nature, while tying our mind firmly (*musubi*), we reinforce our life. This is the essence of the spiritual homecoming that rejuvenates our Japanese people.

Poly-gods and mono-god

Japanese myth teaches that *Ame-no mi-naka-nushi*, a supreme omnipotent deity, lived in the center of the Universe. As creation's activities developed, he named deities, one after the other, according to their duties. Finally there were eight million deities. These eight million deities combined become one deity, *Ameno minakanushi*.

Gods' names are given to the Sun, the Moon, and stars. On Earth, these names are given to mountains, seas, rivers, soil, water, trees, rocks, stones and other gifts of Nature, as well as to food, even to our excreta. From this naming, we can learn how much people in those days respected Nature and paid careful attention to sustaining it. It can also teach us the importance of leaving human life in the hands of Life itself and of Nature. It can teach us the value of becoming worthy of living in harmony with Nature by developing the gentle consciousness which used to be intrinsic to Japanese culture.

When the hand of the Universe is open there are eight million gods. When it is closed there is one God. One God is at the same time many gods. Many gods at the same time are one God. The Western rational mind may find it difficult to understand this kind of Japanese thinking. The Shinto deities seem to bear little resemblance to the gods of monotheistic cults. When we observe Nature and Earth's ecosystem, so diverse, yet so carefully related, we see that truth may appear in paradox and that both the rational and the non-rational are realities within the Universe as a whole.

There are many different professions. Roles played may differ according to profession, and professions may change according to the roles. Likewise, God carries different names according to the role. Many gods can be one, and vice versa. This description may appear to be inconsistent to the Western mind, but Japanese have accepted it without hesitation. It is not inconsistent within their consciousness. That is why Japanese may observe Christmas, then visit traditional shrines for worship as the New Year enters.

Japan is said to be a *furoshiki* culture. The origin of *furoshiki* lies in the mono-god-poly-gods idea. *Furoshiki* is just a plain cloth. It is soft, so its shape can vary according to the materials wrapped. It is very flexible and free. It can be round, square, or triangular; large, medium, or small. When there is nothing to wrap, it can be folded and put into a pocket. While appearing to be nothing, it can easily wrap anything. A bag cannot do this. Since its shape is fixed, it cannot accommodate everything. Instead, one shape of bag may need to be switched with another according to what is to be put inside. The bag may be independent but it is not flexible.

Japan is also a country of "*chijimi,*" shrinking culture. Life with *tatami*, a mattress made of a special grass, requires that everything be folded and put in order. This applies not only to *kimono* (a Japanese traditional robe), low tables, and *futon* (Japanese bedding). People sit on the floor with their knees folded.

The same room gets different names according to the uses it may serve. For instance, it becomes a dining room at mealtime, a guest room when a visitor comes, and a sleeping room at night when the *futon* is set out. Because of simplicity of these everyday life customs, easy-to-carry and compact items proliferate. They make life more convenient. The Japanese tradition of putting away and putting out (hold-and-spread) household items daily is reflected in the dexterity of their hands.

Agriculture as musubi of Heaven and Earth

I described earlier *musubi*, the function of combining or uniting. Observe *musubi* in agriculture. *Musubi* is the principal factor for human life. Since ancient times, the idea that "agriculture is the basis of a nation" has been part of Japanese consciousness. Its origin lies in myth and the idea has been passed down through the generations. Today, however, too few people pay attention to the spirit of *musubi* which used to be inherent in agriculture. The current commercialization of rice and other agricultural products seems to be the result of the *musubi* spirit being forgotten and, as a result, the farming mind has been diminished. Japanese agriculture today is being ruined by farmers focusing their minds only on business and by society advocating economics as the priority for life. Productive land is ruined. Worse, peoples' minds are ruined. It is really a seismic change of consciousness.

The introduction of a rice culture into Japan is a comparably recent event, said to have occurred about 2,300 years ago. Japanese mythology teaches that the rice plant was bestowed by Heaven and, as such, was sacred. Ever since the birth of our country, agriculture has been at the center of progress and our traditions, and our indigenous culture has been built around it. Farmers were called *oomi-takara*, the treasure of the nation, and until recently, they ranked in respect above those in commerce and industry.

Traditionally inherited *musubi* was the function of *tenchi no musubi*, combining heaven and earth. Agriculture was sacred work, transmitting life to all creatures. Farmers were faithful intermediary agents for agricultural production. Although their labor was very hard, farmers realized that human strength was nothing in comparison with *tenchi no musubi*, the Life Chain of Nature, so they nurtured a spirit of prayer, gratefulness and industriousness. At a subconscious level, this spirit has been inherited by all Japanese. Subconsciously we each recognize there is no place for human ego or ideology in the continuing and eternal flow of the Life Chain. Well-devised agricultural methods have been based on recycling and on introducing natural fertilizers that do not harm the soil or the farm produce.

More recently, however, scientific advances have paved the way for increased agricultural production. The application of synthesized fertilizers now plays an important role. Soil science, fertilizer science, cultural science, crop physiology, agricultural meteorology and other narrow disciplines have emerged until it seems that no more specialization is possible. The proliferation of disciplines have given the impression that agriculture has made authentic advances.

As the proverb, "You can't see the forest for the trees", indicates, we should not lose sight of the fact that the processes of agriculture, our basic industry, are fundamental to the very functioning of Heaven and Earth. Continually expanding our agronomy is leading agriculture into a holocaust. The time has come to once again pay attention to our origin. Whether we be politicians, scholars, producers, consumers, or ordinary citizens, we must now insist that agriculture, indeed industry as a whole, embrace the universal principles of Nature.

Biotechnology is no exception. Each of the bits which scientists manipulate is part of the Life Chain and Cycle. Each operates according to the principles of Nature. While attempting to manipulate the composition and decomposition of organic substances and organisms and the scientific transformation of inorganic materials, we must continuously be aware of the real nature of the Life Force of the Universe so that the agriculture that we develop once again can serve as a source of fundamental energy.

Today, the soil is tired, injured and sick. Even if straw is applied to paddy fields as an organic substance, it doesn't decompose well because of the rapid decrease of microorganisms. More and more fields are becoming so hardened that ordinary farming machines can barely till the land. Will we choose to rejuvenate these fields? Where and how will life be sustained if the land becomes unproductive? It is not only the land whose fertility is declining as a result of modern farming. Rivers are becoming polluted as a result of the excessive use of chemical fertilizers. As a result, agricultural products no longer contain the energy generated by *musubi*, the combination of Heaven and Earth. The volume of nutrients then declines. The food we eat—grain, vegetables and fruit—retains its shape, color and taste, but it no longer carries the fundamental energy we need.

The area of forestation is decreasing all over the world; deserts are expanding. These are warning. Land is nourished by the creative activities of Nature. Where the life forces of nature flow, organisms live and human beings settle down. Historically, human need has generated human industry and the land has provided us with bounteous blessings. Thus, human endeavor has influenced the direction of the flow of Life itself.

As each of us is dependent on the flow of Life for our—and our children's—survival, we each have a vested interest in the endeavors of others when their actions can affect our very survival. Thus there is a strong argument for collective constraints on all human industry.

Agriculture is no exception since it represents *musubi*. The spirit of agriculture emerges in other industries. *Musubi* spirit originates in the combination of Heaven and Earth and, therefore, flows with the Life Chain and Cycle of the Universe. Agriculture, fishery and forestry—dealing with seas, rivers, mountains and plains—are direct recipients of and participants in the Life of the Universe. From this perspective, industries such as mining that contribute to human prosperity have their origin in agriculture.

Minerals, resources accumulated underground, are the products of Earth's creative functions. Fossil fuels such as oil and coal—the remains of ancient organisms—are the products of *musubi* created by Life on Earth. These natural resources are God-given, produced by the *musubi* of life on Earth and in the Universe, perpetually connecting past, present, and future. For human life to continue to sustain itself with this blessing, we must exercise moderation in accord with the principle of Life Chain and Cycle. God-given common sense urges us to be moderate and modest as a matter of course. We must cease plundering and aggressively attacking Nature and our God-given resources. They are for the benefit of all the peoples of the world, not just for the present generation. Future generations also will inherit the benefits of the Life Chain.

In 1993, one of the worst years for rice production in half a century, Japan had to import 2.3 million tonnes of rice as an emergency measure. Japan's rice imports may well have adversely affected the global market. We Japanese must reflect on our excessive diet and unnatural agricultural practices. We must consider the lives of people who do not have easy access to even one grain of rice. With plenty of money we can buy food abroad. Imports seem like an easy solution. Yet, we must ask, is this the best way?

With limited land area for agriculture, Japan has had to reclaim hilly lands. Today, at last, Japan has the capacity to produce enough food grains for its own needs. Agricultural skills and knowledge have improved. Synthetic fertilizers and other chemicals have developed. Yet, our attempts to further increase production have put an unreasonable burden on the land, stretching it beyond its ability to produce. When the increases were attained, we began to reduce the land area under cultivation, failing to recognize that the increase was the result of excessive chemical use and overproduction of the land. These agricultural practices, based on selfish motives, indicate the extent to which we have lost the sense of gratefulness for the blessings of the land.

The freakish weather that led to a decrease of rice production helped to mask the destructive role that chemicals, synthetic fertilizers, and even large farming machines have played. We must urgently recognize that the most important elements for growing rice are soil, water and solar energy—the mighty forces of Nature. Bumper harvests come when human labor is directed in harmony with these natural forces.

We must remember that Earth made soil through long years of evolving natural activities. It created an environment capable of sustaining all living beings. Agriculture, when it is guided only by economic motivation and is forgetful of the blessings of Nature and Earth, can totally destroy the means upon which human survival depends. Agriculture is our means of producing the food necessary to sustain Life. It is totally out of the question to demand excessive production of agriculture to maintain its competitiveness, as has become our habit with other industries. Agriculture is fundamentally different from other industries. Excessive productivity, forced on agriculture, will destroy the natural equilibrium of land and, ultimately, the land will lose its ability to produce. The land will become barren. It may not recover. So, the selfish motives of people are a profanity against Nature. Genuine farmers have different motivations.

Policies reducing the amount of land set aside for cultivation are advocated by people who have no understanding of the principles of Nature or the global environment. Land, with a life of its own, has limited patience for the selfish motives of we humans. It may soon face us with an ultimatum for our rebellious acts.

CHAPTER FOUR

The Truth of Life Chain and Cycle

The Universe is full of 'spiritons'

Dr. Yonosuke Nakano, the founder of the OISCA movement (and my father) devoted the second half of his life to studying *shinto*. He offered the name "*shinrei genshi*" to *kami* that is described in *shinto* as "*taka-amahara ni kami tsumari masu*," or, deities are assembling in the Universe. He often spoke as follows:

> Although the universe is expanding limitlessly and endlessly, it is not an empty space, but filled with deities throughout. These deities are active. When they are observed in *uchu-reikai*, the spiritual sphere of the Universe, they are *reishi*, mini-spirit or quantum of spirit, that can be seen or heard only by spiritually enlightened eyes and ears. I call them *shinrei genshi*, divine spiritual atoms or '*spiritons*'....
>
> These spiritual atoms are functioning in a sphere smaller than the micro-world of science that recognizes atoms and elementary particles. Their functions are so delicate, without beginning or end, limitlessly large, endlessly small, extremely pure and more dignified and supremely perfect than any human words can elaborate.

Earth, accommodating human beings and all other creatures, is only one of the planets in the solar system. The galaxy that includes the solar system, having about 100,000 light-years' diameter, is a big celestial body composed of about two hundred billion fixed stars, gas and dust. More than two hundred billion such galaxies are said to exist in the Universe. No human words can accurately describe such scale. Time is eternal and infinite for the Universe.

From where comes the force that transmits life to so limitless a number of celestial bodies and gives the energy to the life and death cycle of stars? It must live in the vast space of the Universe, if my father is right; and I believe he is. The Universe is filled by *shinrei genshi*, or spiritual atoms, and is itself Life. These spiritual atoms function continuously without even a momentary recess. They are part of and sustain the life of every living thing, from the extremely large (celestial bodies, Earth) to the extremely small (bacteria and microorganisms)—including human beings, animals and plants. Each flows with Life itself. None can survive if it strays from the greater Life.

Flow of Life—Will surpassing human intellect

Yonosuke Nakano also proposed:

> When a sign of origin appeared in the Universe, there should have been a will for creation. The will of the Universe exists, but it does not move. When it creates a flow with an objective, it is transformed into a great spirit.

Grand scale creative activities carried out ever since the origin of the Universe are the great industry of the Universe. They have no end and are far beyond the current capacity of human knowledge and imagination. In the Universe the life and death drama, creation and collapse repeat endlessly. I cannot conceive of this being purposeless. Yet, I do not personally know the purpose and it may be beyond human knowledge. Nevertheless, I am confident of the existence of a great Will in the Universe, administering every aspect of its functioning.

What, then, is the objective of the Will of the Universe? Can it be to create living things and to have the Life flow eternally? That process began as soon as the big bang took place. Life and death repeat in the Universe, in human life and in all other living things. This is so today and will be so in the future—eternally. Life will continue to flow, according to the will of the Universe, toward the eternal future. The industry of the Universe continuously occurs in extremely large and extremely small spheres. It promotes creation and advancement at every corner, and it associates with all creatures in the Universe. Its function never stops. It goes on creating life. It is grand art itself.

In the past, scientists regarded the universe as chaos. They began to recognize that order actually existed after Newton discovered the law of gravity. Chaos theory is beginning to emerge again. Science books now explain that chaos has veritable laws and order.

Eternity of life

The common understanding of "life" applies only to physical living beings on Earth. However, life is flowing ceaselessly in the Earth as a whole and in the Universe at large. A dead human body, for instance, regardless of the type of burial practice, decomposes as bacteria and microorganisms help it return to the land. Thus, a human life is not for the self alone, nor even for human society, but remains eternal, even without form, in its living micro-elements.

The essence of life is more than its physical component. It also includes a spiritual component that we recognize as its eternal "soul". Eternity is the link with the great spirit and the life of the Universe. Our ancient forerunners called a human being *hito*, an agent to hold a soul. Human advancement requires elevation of its spiritual as well as material qualities.

It is important for us each to recognize how valuable and blessed it is to be given life on Earth to have a soul as well as a physical body. We should humbly remember that, before any life is born, the law and order of the greater Life of the Universe existed. From this perspective we can begin to see that there are rules of life that are to be faithfully observed by every human being.

Selfishness and tyranny of human beings

Science and technology are providing the means that bring to each of us what we believe to be a faster and more convenient life. In our ignorance we do not realize that what we believe to be sweet drugs may actually be poisonous. As a result of endless human pursuits in the name of human progress, Nature is being injured and degraded. The deteriorating global environment may soon be irreparable. How will the human race continue to survive, alone in this deteriorating environment which it is creating?

What will be the fate of the other organisms that have played their role in supporting human life? Throughout the ages, organisms have evolved according to Nature. How long did an elephant take to get such a long nose, a giraffe to get such a long neck, and a turtle to get such a hard shell? Earth's ecosystem progresses according to evolution. The human race is moving ahead so fast that there is no time to look around.

In the process of evolution, on the one hand, new species emerged; and, on the other, there were species that became extinct due to climatic or geographical changes, or to having lost in the survival race. However, the speed of extinction of species that we are experiencing and facing now is extraordinary. Never before has such desecration been witnessed in our long history. Clearly, the human race is responsible. It is said that 65 million years ago, when dinosaurs lived, species died out at the rate of one in every 1,000 years. The Renaissance of the sixteenth century then became a turning point. Disappearance of species accelerated. The rate increased to one a year in the early 1900's and then to 40-100 species a year in the 1950's. According to one reliable report, 10,000 species disappeared in 1975.

A special research report, "Earth in the Year 2000", was compiled and published early in the 1980's by an advisory committee on environmental issues formed by ex-President Jimmy Carter of the U.S. The report projected that the number of species destroyed during the next two decades will reach 500-600,000. More recently it has been estimated that, on the average, every year from 1990 to the start of the next century, 40-50,000 species may cease to exist

The speed of development today, based on our selfish motivation and the economic priorities of the human race, is bringing us into terminal conflict with our global environment. A great number of lives are lost because we are ignorant of the natural flow of Life and we have little or no consideration for other organisms. Human beings have created their own civilization—one that seeks narrow human advancement and prosperity within their own limited understanding. That understanding, unfortunately, does not extend to the rest of life. We are only tightening the noose around our own necks. Some now realize the impasse and that the unrestrained acts of human beings may bring about the end of life, but they are not yet able to find a way out.

Commercial products and services that stimulate the endless desires of people are flooding the marketplace. We may learn far too soon whether these desirable items will lead to real happiness and wealth. Even though it is likely that they will not bring real happiness or wealth, demand for them increases.

Global population is exploding. As I am writing this message, a watch that I received as a gift from the United Nations is continuing to tick off the global population. It crossed 5,600 million and is fast growing towards 5,700 million. Food production is becoming a serious problem of supply as much as it is a problem for the environment. We may not be able to produce sufficient food to feed the human population. Too much land has already lost its fertility; it may never recover. Since agriculture is so important, fundamental reflection on the universal principle of the Life Chain is essential. Old habits are not easy to change. It will be difficult to get away from the system of economic materialism. Nevertheless the reality in which we live is that the environment of our planet is currently moving to bring the whole life structure to an end.

If human society continues to proceed as we are, we will soon see more and more signs of environmental destruction in many parts of the world. Grain markets will be further confused and freakish weather may continue to occur globally in the years ahead. If diplomatic efforts relating to natural resources and food production remain ineffective, the potential crisis that will ensue will have no comparison with the oil crisis. Agriculture deals with the essential items for life. The likely result is clear. Nations that cannot feed themselves will grow weaker and more vulnerable. Those that can will find they have less of an excess to share with others. This is a path to conflict not unity.

The survival of the species is dependent on the whole of the human race and on our planet as a whole. It requires harmony between the material and the heart. The laws of Nature are exemplified in the Life Chain and Cycle. We should act with that thought in mind. We can no longer afford to avoid changing our behavior. We must remember our place in the Universe, we must reject selfishness and tyranny, and we must take deliberate actions toward a sustainable future.

In June 1992, the United Nations Conference on Environment and Development (UNCED) convened in Rio de Janeiro, Brazil. It was the largest international conference of the present century. Known as the Earth Summit, heads of states, presidents and prime ministers from more than 170 countries came. Many people thought that the conference was already too late. People aware of the probable dangers ahead came from all corners of the world. A common consciousness was achieved by participants from many countries and races, including people from both simple and complex life styles. The Rio experience was a big step forward. The second step, the third step and all the further steps are now essential. The true fate of the whole human race is in our hands.

CHAPTER FIVE

A New Earth Ethics

Just before UNCED convened, a "Sacred Earth Gathering" took place near Rio Centro, the conference center. The participants were not many and the facilities were simple. The environment created by these caring participants from many parts of the world was full of enthusiasm and love for "sacred Earth". I was invited to share at this gathering the idea of "Earth Ethics" that I have been advocating.[11]

The dawn of a new era

In the past, many prominent people have advanced thoughts on ethics. They have taught that we ought to be loyal to our nation, to be thoughtful to our parents, to respect our teachers, to help people in trouble, not to lie, to use our belongings with care, and to make the most of them. These moral codes are important and necessary to build a comfortable and progressive society. Ethics may differ somewhat from country to country depending on climate, natural features, religions and cultures. Nevertheless human beings have always been at the center of these teaching.

---

[11] The following is edited from an address by the author on the occasion of the Sacred Earth Gathering, a parallel meeting of the United Nations Conference on Environment and Development (UNCED), Rio de Janeiro, Brazil, July 1992.

Ethics are important to nurture morality. However, what is happening today on a global scale cannot be dealt with the same guidelines human beings have used to date. Today's problems point to the need for what I call "Earth Ethics."

Human Ethics ought to coincide with Earth Ethics. Unfortunately, our present human society has yet to see the connection. Humanity is still immature. We are not yet sufficiently conscious of the greater laws of Nature, and of the wider life cycles and of the great unifying force of Life that flows through all living things and extends throughout the Universe. Human beings can find the strength to enrich their hearts and minds from the source of Life itself. We must raise our vision beyond our own immediate environment and our own material desires to focus on the greater whole of which we are each but a part if we are to become consciously aware of the correspondence between ethics for everyday behavior and Earth Ethics. Only by raising our vision can we ultimately enhance the eternal habitability of our beloved Earth.

As symbolized by the words, *fin de siècle*, or the end of an age, many people are now becoming aware that the human race is at a crossroads. Religions that were formerly in the forefront of human consciousness are now unable to grasp today's trends. They stick rigidly to antiquated dogmas and display utter intolerance. They lack the spiritual element and have not been a force to alter the mistaken course of human behavior.

Present society attaches excessive importance to economic efficiency and to satisfying human greed. Much human energy mobilizes itself to produce material goods. Freedom is understood to mean allowing man to freely fulfill his desires. The net result is that people strive to increase their material consumption. The freedom of mass consumption has captivated modern man. Intensified competition, mass production, mass consumption, and disposal of enormous wastes repeat themselves in a vicious and destructive circle. The effects extend throughout the world. Because of the never-ending increases in production, the resources of Earth are rapidly being used up. We are now being forced by the life threatening state of the environment to stop and reflect.

Response to rising population and poverty

Increasing burdens on the ever deteriorating global environment arise from the population explosion and worsening poverty. In April 1992, the United Nations Fund for Population Activities (UNFPA) made public the World Population White Paper entitled, "Toward Harmony with the Earth." This document starts out with the fact that the population of the world had reached 5.48 billion. It projected that, in the following ten years, increases will be recorded at the rate of three persons per second, or 97 million people per year. At this rate, the world's population will surpass the ten billion mark by the year 2050. Thirty percent of the increase will occur in Africa and twenty percent in south Asia. How can this single planet with its limited resources sustain such an increase in population? It can comfortably sustain two or three billion people. When the population reaches ten billion, and if each of us was to consume at the level of the average U.S. citizen, current sources of crude oil would be exhausted within five years. The lack of arable land would drive environmental refugees to the richer countries. The number of people below the absolute poverty line already exceeds 1.2 billion. They could all become refugees.

This disparity among peoples poses a serious challenge to the entire human community. Unless the gap between the rich and the poor narrows, the potential for conflict will remain high. But the task before us is much greater. Even without conflict, we face an uphill task simply to promote sustainable development and to enhance the environment.

We begin that uphill task by recognizing that human happiness and satisfaction cannot be derived from material prosperity alone. If we can attain total happiness with material wealth, why do vast numbers of people in developed countries, despite their affluence, feel such anxiety and stress? Human life is both material and spiritual. Material prosperity cannot lead to real happiness and peace of mind in the absence of enrichment of spirit. I do not mean to overemphasize the spiritual or to make light of material progress. However, the abundance of drug addicts, crime, domestic discord, divorce, and juvenile delinquency in developed countries is a clear sign of spiritual poverty. Even in developing countries where there is a great deal of material poverty, you will find rich and warm hearts, sweet smiles, kindness and helpfulness to neighbors and relatives, particularly in rural communities. Such warmth in people, however, does appear to diminish as you near the bigger cities, where human feelings for other people seem to be drying up. Perhaps this is a result of people in bigger cities being separated from Nature by the very conditions of city living.

Scientific advances

Today one cannot look at world events without being influenced by science and technology. TV and other mass communication networks have enabled us instantly to transmit and receive information about natural disasters, about great achievements and about how other people live.

Such information networks helped to remove the wall that blocked the free movement of people to and from the former communist countries. Without the contributions of modern science, thousands of people from all over the world would have been unable to travel to Rio to gather to have talks on the critical issues of environment and development. These same jet aircraft which enabled us to gather in Rio, have powerful engines which consume enormous amounts of fuel and discharge large quantities of toxic emissions as exhaust.

Useful devices developed by science have also enabled man to study nature closely and discover previously unknown life forms and patterns of behavior. Optical instruments now permit man to observe from the macro-cosmos to the micro-world. A recent TV programme showed the eco-systems of a jungle in South America. Surprising scenes illustrated how wonderfully ants live together with numerous plants in the forest. The ants persevere in the face of dangers. Their interactions sometime appear humorous. The technology needed to film a two or three millimeter long ant so clearly with TV cameras was also impressive. We could see a group of ants destroy and carry away the residue of dead plants and animals to restore the forest. These extraordinary glimpses into the lives of various animals and plants in the jungle can provide us with much useful information. It is a learning experience that can be much more valuable than school studies, that often tend only to cram fact upon fact into children. The jungle can teach us about life. Its detailed study will involve many subjects: mathematics, science, social science, art and a number of others.

In 1989, Voyager 2, launched by the U.S. space agency, NASA, transmitted from afar beautiful pictures of some of the planets which revolve around the Sun. Those photos of Uranus and Neptune were marvelous and unforgettable. Astronauts have also admired our beautiful blue Earth from outer space. What is it that supports these beautiful planets, makes them rotate and move around the Sun so exquisitely, with faultless precision? Within this Universe, there must undoubtedly be enormous powers and energies at work which are beyond human comprehension. Yet, don't we know this, somehow! Is it not an intrinsic part of our very nature and instinct to adore the Will of the Universe and to seek to conform to that Will. I suspect that in the not so distant future, science will recognize such a notion and it will become part of science. Whether one regards Earth as a life-possessing being or a mere material object will determine one's overall view of the world. I believe the former viewpoint is more relevant to and necessary in the coming age.

The scientific attitude toward the environment is changing. Scientists now do much interdisciplinary research as exemplified by the new field of Earth Life Systems Science. Science is now beginning to recognize that the numerous elements of environmental or ecological systems intermix and intertwine, that they are linked to each other through life chains, that the elements of the whole cannot be separated and that they must be examined in totality, together with their visible and invisible connections. It is a considerable improvement that science now tries to grasp holistic views. In the past, science took the path of specialization. Now its direction appears to be reversing, towards integration. Specialized knowledge in a certain field can be part of promoting the true welfare of humanity, but it must be applied within a wider field of knowledge if it is not to cause harmful side effects. This is quite apparent in the field of medicine. This new change of consciousness is now leading western medicine to incorporate some elements of oriental medicine—truly a great advance.

Water, soil, and forests

Today we know that Earth has a unique place in this Universe. It is the only planet in the solar system with an abundance of water. Oceans and seas account for about 70 percent of the surface. If Earth were a little closer to the Sun, all of its water would evaporate and it would soon be as barren and rocky as Venus. The birth of our Earth is a miracle. Only this miraculous Earth is inhabited by life as we know it and for living things to emerge on Earth, water was necessary and inevitable. The soil, the water and the atmosphere that support humans and all other living things on Earth have developed over a very long geological history.

Earth was born about 4.6 billion years ago. Within this Universe, through dynamic interactions with the Sun, which is the central actor, and other bodies, Earth evolved into its present form. After a considerable passage of time, about one billion years, the precursors of current living things began to appear in the sea.

Before the first plants moved up onto the ground about 400 million years ago, only rocks existed. When the elements of the atmosphere changed, making it suitable for living organisms, plants flourished. Gradually, fertile topsoil formed. One-half of soil is air and water. The remaining half is organic matter (like fallen leaves)and inorganic matter (like sand). As the surface soil grew gradually denser, it began to hold increased amounts of water, supporting more plants and more numerous kinds of organic matter. The existence of plants enabled many kinds of animals to develop and prevail. They all were linked to each other through the food chain and other connections.

The heat of the Sun and the flow of air create the cycles that carry water from the oceans into the atmosphere and down again onto the land. The water then returns to the sea. During this movement, some water remains in the soil, providing a supply of moisture for all living things. Soil, water, atmosphere, sun energy, plants, animals—all mutually interact with one another to form the marvelous eco-systems that provide the balance of the great cycles in Nature.

Soil is composed of inorganic matter such as air, water, particles, and pebbles, and of organic materials like excreta and the residue of plants and animals. Micro-organisms in the soil live on the organic matter. The functioning of microbes decomposes organic matter and turns it into inorganic material. This then dissolves in water where it can be absorbed by plants. The purifying function of the soil is nothing more than the natural activity of microscopic living things. Just recently the Japanese deep-sea explorer "Shinkai 6500" collected soil samples from the depths of the ocean and found that the soil contains bacteria that eat crude oil.

At the surface, on the other hand, a large amount of agricultural chemicals and fertilizers are doing considerable damage to many of the useful organisms in the soil. We destroy the soil at our peril. It cannot be replaced easily or quickly. It takes hundreds of years to form one centimeter of top soil. Worse, the damage to microscopic life threatens the processes that purify the soil and the water.

To keep the increasing world population at even a subsistence level, we will have to deal with more than the problem of food supply. Without ensuring safe water and clean soil, the very foundation of our life will be at stake in the future.

Plants came into being and animals emerged, and life flourished on this Earth. The human race came into being last. Why is it that humans—the newest occupant of this world, only one of five or ten million species—have been given the highest intelligence? Surely human beings are not meant to be the self-centered rulers of this world. Each living creature lives up to its potential as a species. The human race, now dominating the world, is encroaching on the resources of all other life, cutting into mountains, polluting the seas, rivers, atmosphere, and soil. We are doing such devastating harm to our Earth.

The human idea that we are at the center of this world has produced enormous egoism and arrogance. These are the characteristics which have led us to fail to live in harmony with Nature and have encouraged us to ignore the laws of Nature and of Life. I am not suggesting a return to primitivism, nor am I suggesting that we should give up such conveniences as automobiles, refrigerators, washing machines or vacuum cleaners.

Human beings, who presume themselves to be the lords of all creatures because of their gift of higher intellect, must now earn the love of all these other creatures. We can no longer ignore the visible and invisible gifts from other forms of life on this Earth. We are dependent on them. We do not have the right to interrupt the life and food chains in Nature, pollute them or arrogantly use such gifts once and then throw them away. The Life Cycle does not end with human beings. We must repay all beings for the gifts that they have provided and continue to provide. We must learn to use the superior intellect we claim and learn to live in harmony with Nature. We must find the wisdom to create a more prosperous and glorious world on Earth. Not until this is realized will mankind feel the happiness of the perfection of humanity. It must be the mission of humanity that man helps all other beings fulfill their respective objectives while coexisting peacefully with them in a harmonious and amicable world. Man should cease immediately the burning and bulldozing of the forests and the destruction of the soil solely for his own convenience.

A lesson from the past

The downfall of some ancient civilizations, such as the Mayan and others, can, to some extent, be measured by the devastation of their forests. As their populations increased, these people expanded their farming fields. They constructed boats and buildings and engaged in warfare, all of which consumed large quantities of wood.

The destruction of the forests resulted in decreased organic matter in the soil. The soil then lost its water preserving capacity. When it rained, erosion followed. The land then became desert, deprived of nutrients. A civilization that strips fertility from the soil must finally pay for its mistake. Nevertheless, man repeats this mistake, today on a global scale.

We must remind ourselves that Earth is alive and active. It is part of the great life of the Universe, working together with it and with all of its components, from microscopic entities to the other celestial bodies. We have forgotten to pay due respect to Heaven and to Earth. Nature is at times stern and awe-inspiring. It is, however, at the same time, so dear to us human beings. Today human consciousness is so clouded by material progress that we have lost sight of too many important aspects of life. In this turbid flow, we must now stop and stand still for a while to reflect upon our Earth and to listen to the voice of Nature. The solution to global problems will depend solely on us changing our consciousness and on our ability to assume a new attitude. We will achieve our goal only if each one of us individually can change his or her consciousness and do what is required.

I am often told that the leaders of native groups in the Amazon and indigenous people elsewhere are opposed to the development of the forests where they live. We must listen to their voices. For generations, these forest dwellers, without relying on scientific devices, have lived in harmony with Nature, have earned their livelihood and, above all, have protected their forests. They should be fully consulted and their lifestyles should be fully respected before any development scheme for their forests is carried out.

Native Americans refused to join in the celebrations of the quincentenary of the voyage of Columbus which Europeans celebrated. Discovery of the American continents can be called the dawn of modern civilization. It was also the start of 500 years of destruction of the traditions of the native people. Their traditions and way of life were suppressed by the foreign cultures that overtook them. It is not surprising that their few remaining descendants refused to join in the celebrations. If you study world history, you will notice that many cultures and civilizations have been annihilated through foreign domination. Such losses are, indeed, heart-rending. Because of the lingering after-effects, many aboriginal peoples are still not able to regain their full identities. There is substance in the claim that the prosperity of advanced countries is based on the sacrifice of other nations.

An idea whose time has come

Looking back over recorded history and judging whether human actions have been good or bad will not of itself produce solutions for the present problems. Emotion would only conceal truth. What we should do is to return to the starting point. We must consider what human beings are and what human purpose is. We now stand on the threshold of the twenty-first century. We are not only at the end of a century but the end of an era in human history. We must renew our spiritual consciousness. It is often asserted that the industrial revolution was the beginning of our mistake. I think that we must search our souls and look much further back.

If we don't, some remarkable innovation may well lead us to believe that a partial solution is possible through technological advance and we will soon forget the dangers implicit in our very way of life and return to our habits of excess consumption and wastefulness. Many will think that the problem has been solved. Others will think "What can I do? An insignificant person like me cannot contribute enough. Any action I take will make no difference at all." A wise man once said "Danger is past, God is forgotten."

It is incumbent upon us all to think over the past and to recognize that our long human history has been filled with the law of the jungle, insatiable greed, exploitation of others, and abuses of the resources of the Earth. It is essential that we should try to remove self-centeredness from our minds. Each of us must concentrate on giving priority to the whole rather than to individuals, on trying to achieve harmony instead of confrontation, cooperation instead of conflict, and amity instead of domination. We must recognize that all living beings on Earth, including animals and plants and micro-organisms are linked together by the mutually supporting intricate Life Chain.

In 1992, the International Union for Conservation of Nature and Natural Resources (IUCN), the United Nations Environment Programme (UNEP), and the World Wildlife Fund (WWF) jointly issued an updated *World Conservation Strategy*.

This strategy is founded on the conviction that people can alter their behavior when they see that it **will** make things better; and can work together when they need to. It is aimed at change because values, economies and societies different from most that prevail today are needed if we are to care for the Earth and build a better quality of life for all.[12]

If people truly realized the dangers creeping toward them, those who have been greedily devouring material wealth would cut back. That is not to say that we should immediately return to the life style of 50 to 100 years ago. Certainly, as a first step, cutting back to the levels of only five or six years would not be so difficult. However, we must go further. We must go to the point of sharing the pain. It is not realistic, of course, to totally give up economic development in the face of a dramatically increasing world population. Development should be carried forward where essential to human survival. However, development that is aimed only at satisfying covetousness and enjoying the conveniences of affluent society can no longer be considered acceptable. If we human beings wish to seek a richer life, we should seek richness in another aspect of life, in our spiritual aspect.

Today, many leading companies are exerting their utmost efforts to manufacture products that will impose the minimum of burdens on the environment. Nevertheless, these "environmentally safe" products will consume energy and resources. Unless we are willing to restrain consumption, there will be no change in the rate of damage to the environment, no matter how environmentally sound the

---

[12] *Caring for the Earth—A Strategy for Sustainable Living*, IUCN, UNEP, WWF, 1992.

products may be. What is required of us is a change of consciousness. The old proverb says, "Be content with what you have". These words are profound. The New World Environmental Conservation Strategy defines the following guidelines as the basic principle to realize a sustainable lifestyle: "Respecting all lives in Nature, treasuring them, improving the quality of life of people, conserving the vigor and diversity of Earth, minimizing the consumption of non-renewable resources, controlling human activities within the life supporting capacity of Earth, changing lifestyles of individuals, helping rural communities take measures to protect their environments, creating a national framework to integrate development and environmental conservation, and establishing a global system of cooperation." I applaud the strategy. The remedy rests with human consciousness. Without increasing awareness, nothing really tangible will happen.

I can only repeat, we human beings are nothing but a small part of the Universe. Our survival lies in behavior which is in harmony and co-exists with Nature. We must recognize our own rightful place. We must develop a mind that knows what is reasonable. Together we must reflect on the consequences of our modern life style. This is what the "Earth Ethics" demands.

Appreciation

Almost two years have passed since the "Earth Summit" was held in Rio de Janeiro. The United Nations Conference on Environment and Development (UNCED) was said to be the largest international conference since the dawn of history and also in this century. When I reflect on the global-scale environmental issues that threaten the very existence of human beings, I think that the conference was a timely—indeed epochal—event. With the conference as a turning point, I believe that more people became aware of the global environment and to question "how it should be". Some new kinds of actions started.

Unfortunately, increasingly, the environment deteriorates. Our consciousness and actions are still inadequate to address the situation. Today, I even sense retreat on discussing the issues. "Danger is past; God is forgotten." Since I am acutely concerned about these circumstances, I have published this book, wishing that it can be of help. By reading about the foundation of my way of thinking—which is also a background for OISCA-International's activities—I hope you have been stimulated by new ways of thinking about and addressing the global environmental issues.

In July last year, OISCA had the honor to be the first NGO to be awarded the "Earth Summit Award". This award was created for the commemoration of the first anniversary of the "Earth Summit". On behalf of OISCA, I visited the UN headquarters in New York for the award ceremony. At that time, UN officials and representatives of various circles of the U.S. and Europe urged me to communicate my way of thinking more widely. Because of their encouragement, and my own sense of duty, I have written this book. I have experienced difficulty in expressing what I think with written words, but I will be more than happy if you have been able to appreciate even a part of my ideas.

When I was writing this book, Professor Tsuyoshi Nara of the Tokyo University of Foreign Studies gave me various valuable advice. Also, Messrs. Masami Gomi and Hiroshige Makita of the Astronomical Department, International Foundation for Cultural Harmony, kindly collected data concerning today's astronomical science. I would like to express my sincere appreciation and gratitude to Messrs. Tadashi Watanabe, Akio Tabata, and Fumio Kitsuki who have helped me in translation and proofreading, as well as Messrs. Norbert Netzer, John Tomlinson, and Eric Waldbaum who edited the English-language text and made efforts to clarify my ideas for English-speaking readers.

Index

Aborigines 61

Agriculture i-iii, 73-75, 77-79, 86

Apollo 4, 27-29

Big bang 6-7, 11, 13, 51-52, 54, 69, 83

Children ii, vii, 22, 27, 36, 45-47, 93

Communication i, 38, 92

Community ii-iii, xi, 24-26, 28, 30, 39, 41, 68

Cooperation i-iii, v, x, xii, 65, 100, 102

Creation 7-9, 11, 13, 18, 45, 51-53, 71, 82-83

Culture xi, 44, 59-62, 64-65, 67, 69, 72-74, 89, 99

Dark matter 15-17, 43

Development i-iii, 22, 26, 28-31, 85, 92-93, 99, 101-102

DNA viii, 42-44, 49, 55-56

Earth Life Systems Science 94

Earth Summit 87

Earth Summit Award i

Ecosystem ix-x, 21-23, 72, 85

INDEX

Environment v-vii, ix-xi, 4, 17, 20-21, 23-26, 30-31, 37-40, 50, 55, 78-79, 84-87, 89, 91-94, 100-102

Faith 61

Forest i-ii, xi, 4, 39, 44, 57, 59, 61, 75, 77, 93, 95, 98, 99

Fossil vi, 19, 28-29, 77

Gamow, George 6

Guth, Alan 7

Harmony v-vi, xii, 10-12, 21, 23, 31, 57, 59-60, 72, 78, 87, 91, 97, 99, 100, 102

Hawking, Stephen W. 8, 62-63

Hubble 9, 12-13, 54

International Space Year 2

Japan i-iii, v-vi, x-xii, 5, 13, 26, 29, 38, 41, 44-48, 57-62, 64-74, 77-78, 96

Meteorite 1, 4, 5, 17, 70

Mihozeki meteorite 5

Miracle viii, 19, 95

Moon rocks 5, 27-29

Musubi 70-71, 73-74, 76-77

Nakano, Yonosuke 81-82

NASA 1, 6, 16, 93

Nature v, viii, xi-xii, 9-10, 21-24, 31, 43, 45, 46, 48-49, 56-62, 65-66, 71-72, 74-76, 78-79, 84, 87, 90, 92, 96-99, 102

OISCA-International i-iii, v-vi, xiii, 9, 27, 81

INDEX

Oparin, Aleksander 17

Pollution ix, 23

Population 15, 22, 25, 29, 35, 39-42, 50-51, 64, 86, 91, 96, 98, 101

Religion i, xii, 8-9, 61, 63-64, 89-90

Rice ii, 38, 44, 52, 71, 74, 77-78

Science vii-viii, 2, 9-13, 24-26, 30, 32-33, 39, 46-47, 56-58, 62, 64, 69, 75, 81, 83-84, 92-94

Shinkai 6500 96

Shinrei genshi 81, 82

Shinto xii, 58, 60-61, 72, 81

'*Spiritons*' 81

Spiritual i, 10, 14, 23-24, 26, 31-32, 44, 50, 56, 58, 61-62, 64-67, 71, 81-82, 84, 92, 101

Toutatis 1, 5

UNCED 87, 89

United Nations i, iii, 2, 86-87, 91

Vilenkin, Alexander 8

Volunteer ii, xiii

Women iii, 46